아웃 사이드 Out side
Street Furniture design

Concrete
Book Design Co.

발행 : concrete book design
발행인 : 오창수

주소 : 서울시 금천구 가산디지털2로 157 6F

전화 : 02-542-7858
팩스 : 02-542-7857

가격 : 75,000원

아웃 사이드 Out side

Street Furniture design

Concrete
Book Design Co.

목 차

Seat _ 의자 ... 10

Wastebaskets _ 휴지통 ... 104

Fountains _ 분수 ... 126

Light _ 조명 ... 132

Limits _ 경계 ... 170

Planters _ 화분 ... 190

Tree Grids _ 나무 받침 ... 202

Bicycle Racks _ 자전거 거치대 ... 212

Shelters _ 울타리 ... 224

Concrete
Book Design Co.

Introduction

Ever since the beginning of modern civilization, the urbanization of public space has been a prevailing subject and a subject in a constant state of mutation. Besides the structures, buildings and dwellings that serve the full range of possible requirements, cities as well as villages need streets, parks and plazas in which the inhabitants can move, rest, play, or get together. In fact, many urban design theorists consider that the life-quality of a city's inhabitants depends on the proportion of shared space available and how it is used.

There is no doubt that the urban furniture plays a leading role in making people's presence in these areas more frequent, agreeable and comfortable. A well distributed arrangement of benches, street lamps or shade will contribute to making a plaza more accessible, encouraging a more positive and lively relation between its visitors. Thus, an obvious improvement will ensue, strengthening the social links throughout the neighborhood.

Although many of us only hurry through these spaces briefly on our way to somewhere else, hardly noticing the urban furniture, these accessories are in fact a crucial component, capable of making life much more comfortable. On the other hand, for those who make public space a prolongation of their home, whether they be children, teenagers, retired people or members of any other population group, the lack of these items to which we are all accustomed would be highly traumatic. Hence, in many cities, groups of people have created associations to protest about the neglect of their neighborhoods. There are many housing estates in which local authorities have ignored these requests and the situation of neglect has become extremely chronic compared to other more fortunate districts of the same city.

Indeed, urban furniture seems to have become vital for contemporary cities. Not only can these items become a distinctive feature of a neighborhood and of a town, like London's telephone boxes, but it is their design and the degree to which they address the solution to everyday problems that proves their validity. The adaptation to different spaces, the durability or resistance to harsh atmospheric conditions or vandalism, an item's versatility or multiple use and of course its aesthetic value that are a source of inspiration to architects and designers, who have carried out their projects within the public space of the streets, squares and parks of our urban environments.

In this book we have compiled a selection of the most recent and significant contributions to this field of design. Effective and contemporary urban furniture, with designs ranging from the most classical to the most experimental and futuristic, in form as in the implementation of new materials and finishes. Each item is completely illustrated with photographs and plans, together with a full explanation of the constructive details, dimensions and weight, the installation process and the maintenance required. It is all perfectly ordered according to the functions fulfilled, and the clear and attractive layout with a file-like presentation make it ideally easy and practical to consult. We believe this book to be a good alternative to the usual collection of catalogues, providing an interesting source of inspiration for Town Halls, municipalities, town planners, architects, builders and students alike, all of whom will have access to the contact data of the makers of the outstanding items offered in this volume.

Seats

Baf

Design: **Joan Forgas**   Production: **Alis**

The Baf bench consists of three pieces and optional armrests. The back is made of two boards of Bolondo wood, the dimensions of which may vary according to the number of armrests the bench has. The armrests consist of two boards of Bolondo or copper-treated pine measuring 540x175x50 mm, fixed to the structure with metal bolts. The seat consists of three horizontal boards of Bolondo or copper-treated pine, measuring 2000x175x50 mm. A vertically placed board conceals the edge and performs as a drip. The supporting structure is made of 6mm thick, laser-cut, galvanized steel plate, soldered and with an epoxy powder-coated finish 80 micros thick. The colour is "sable noir". The bench is fixed to the ground with special bolts and security plugs, installed previously.

Banca

Design: **Bernardo Gómez-Pimienta**   Production: **Alis**

There are two versions of Banca, a bench and a chair. The boards in both cases are of solid Bolondo wood or copper-treated pine, measuring 1610x90x40 mm. The back consists of four boards and the seat consists of six. The ends of the boards are held by U-shaped galvanized steel structure measuring 50x50x4 mm, with a powder-coating of epoxy 80 micros thick. The colour is "sable noir". The boards are fixed in place with countersunk bolts.

Vivanti

Design: **Max Wehberg**   Production: **Westeifel Werke**

The Vivanti bench for the elderly provides more comfort and makes standing up easier due to its ergonomically designed backrest, armrests and footrest. Parking your own walking frame in the fitted parking facility creates a seat with a backrest. The Vivanti products can create a good walking route, for example from a residential area to a shopping centre, enhancing elderly people's mobility. The Vivanti benches are designed to create good resting places for the older shopper or walker. Both sides of the Vivanti bench can be put to use and it has supportive armrests and footrests.

Very suitable for a shopping street, the standing support with sturdy handgrips has been specially developed for use in locations with reduced space.

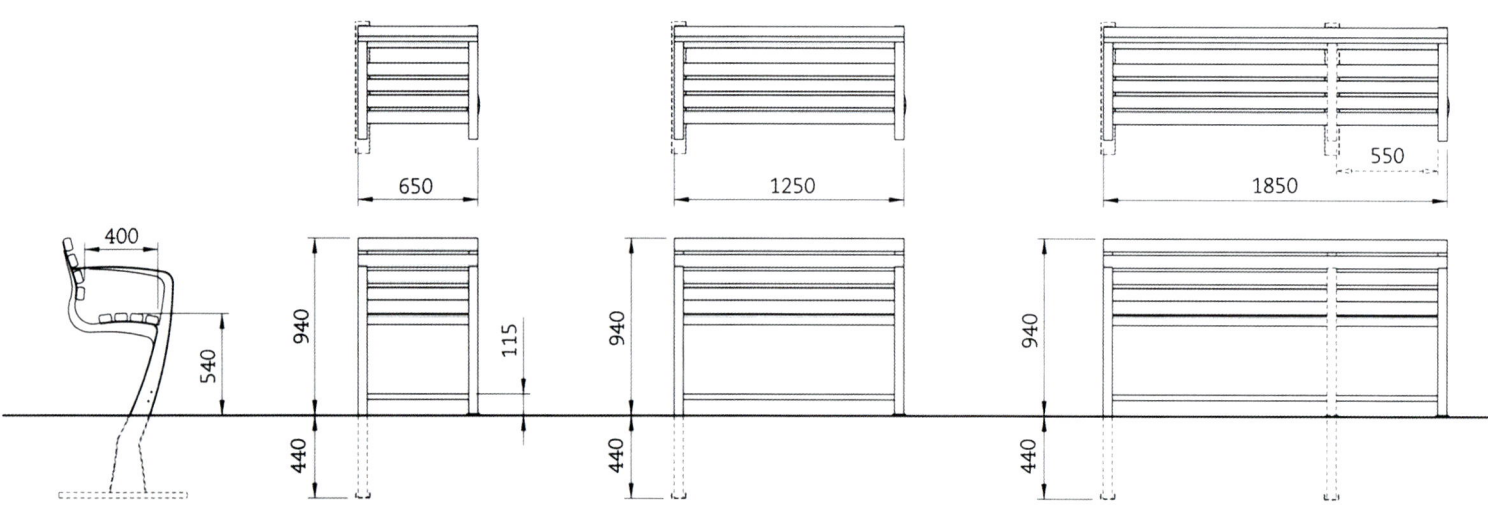

14   Seats

Seats 15

Ameba

Design: **Atelier Mendini**   Production: **Ghisamestieri**

This range of Park benches consists of lateral frames in cast-iron that support alternatively twenty-nine or sixteen Iroko wooden laths. These are impregnated with protective fungicides and are available in two different lengths (1500 mm and 2000 mm); steel ties allow for fastening. The laths may be painted in the colour requested. There is a circular base-plate at the end of the cast-iron lateral components allowing the bench to be fastened to the ground with expansion bolts. The items are delivered separately, for the simple assembly operation to be carried out by the customer.

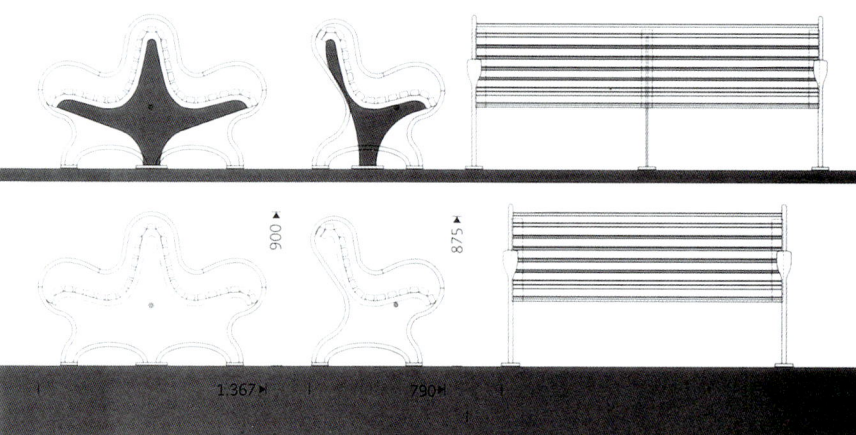

Arona

Design: **Enrico Marforio**   Production: **Ghisamestieri**

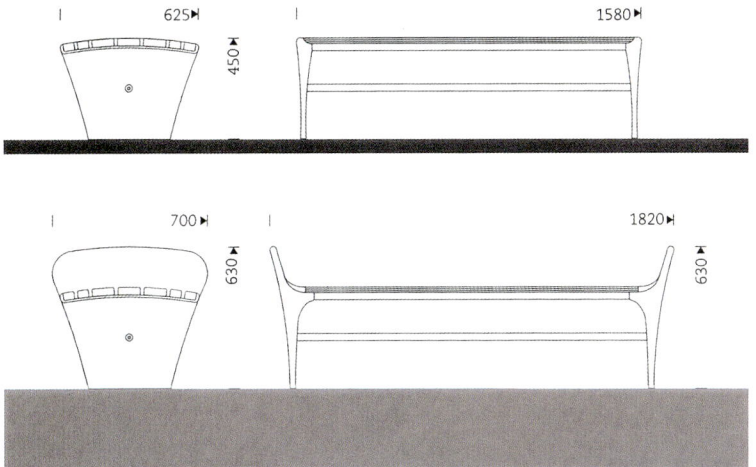

Park benches consisting of two lateral frames in cast-iron and seven Iroko wood laths (differentiated sections 50 × 33 × 1500 and 100 × 33 × 1500), impregnated with protective fungicides; steel fastening ties allow the parts to be fixed together. The laths can be supplied in other colours on demand. Semi-circular feet at the end of the cast-iron lateral components allow the bench to be fastened to the ground by means of expansion bolts. The bench is delivered disassembled and the customer carries out the simple assembly operations.

Flor
Design: **Mansilla+Tuñón**   Production: **Escofet**

Flor is based on research regarding the concepts of equality and diversity that are present throughout the work of Mansilla + Tuñón. In the case of this concrete "flower" they have implemented a subtle game of similitude and difference. The biomorphic and radial form permits it to be used by individuals or by couples with no loss of intimacy.

Flor has been installed in the courtyard of the Archivo Regional de la Comunidad de Madrid, situated in a complex architectural ambiance of different periods, styles, materials and uses. The "flowers" complete this public meeting space that welcomes visitors to the various institutions that share the location.

Bilbao

Design: **Josep Muxart**   Production: **Escofet**

The Bilbao bench is one of a series of urban elements that come under the same name. These reinforced concrete seats have been designed to transmit a feeling of organic softness to the rigidity of the actual material. This has been achieved by using a composition of rounded forms and warped planes. The result achieved is a series of disquieting pieces that seem to be undergoing a contained state of movement. The dimensions have been kept limited to permit a more amiable interaction with the environment.

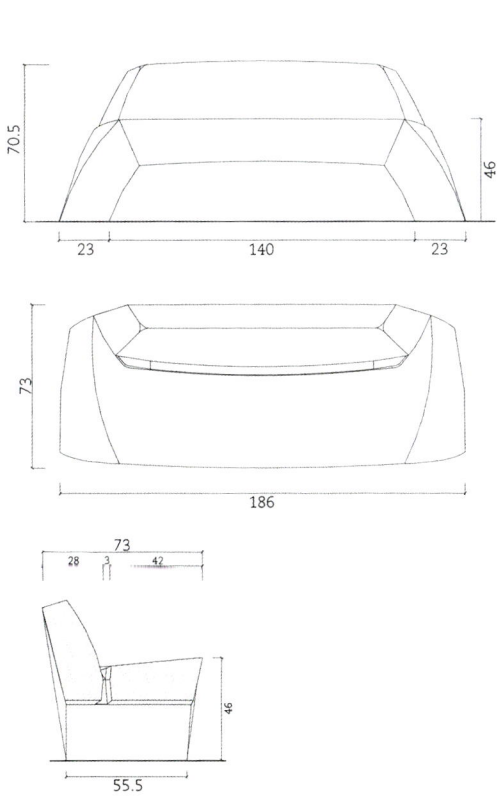

Elemental
Design: **Juan Sádaba**   Production: **Onn Outside**

Elementary in its solid shape and conceptual simplicity, this bench leaves all the options open as to how it is positioned. The independent blocks can be grouped to allow the users to create their own unique formation.

Leichtgewicht

Design: **Thesevenhints** Production: **Miramondo**

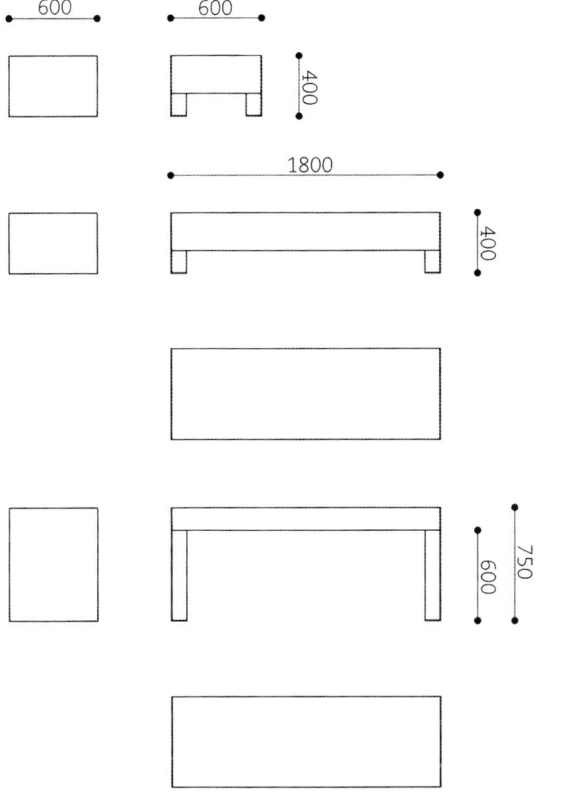

"Leichtgewicht" - visually strong yet light in weight. At exterior construction sites, furniture is often the last thing to be put in place and the installation of heavy items of furniture is particularly problematic in areas that are difficult to access. Using modern fibre-concrete in combination with a supporting steel structure, makes the 135 kg, 115 kg or 60 kg of "Leichtgewicht" easy units to handle and can even be used in statically precarious situations such as roof gardens or underground car parks. The slight material thickness of these items has thermal benefits compared to massive concrete and stone pieces - this material gets less cold. The surface of the concrete is free of air inclusions and limestone edges, lending it an appearance of superior quality. The hollow space in "Leichtgewicht" can be provided with a lighting unit.

Seats

Pancarè
Design: **Eugenio Cipollone**   Production: **Alis**

This four-seater bench has been designed by Italian architect Eugenio Cipollone. The seat measures 200x430 mm and consists of structural "T" and "L" shaped elements of steel, measuring 80x80x8 mm, soldered together, leaving an 8mm gap between the planks. The seat and the sides are made of the same structural "T" irons, finished in powder coated epoxy 80 micros thick, colour "sable noir". The Pancaré chair can be fixed by sinking the feet 200 mm in the ground or it can be anchored to the pavement with specific bolts. The version for sinking in the ground can include an "L" or a "T" shaped back. In either case, the separation between this and the rest of the chair is of 100 mm. The seat of the chair can include rectangular Teak or Bolondo wood pieces measuring 550x430x30mm, with longitudinal troughs for drainage. These pieces are held in place by a nylon knob designed to slide along the gaps in the bench.

Riddle Chair

Design: **Jean Nouvel**   Production: **Alis**

Riddle is an armchair designed by the Jean Nouvel workshop. Basically thought out in terms of use in the open air, this piece of furniture can equally well be installed indoors. It is entirely made of cast aluminium. The various parts are held together and onto the structure with security bolts. The chair can gyrate or be fixed according to the characteristics of the location where it will be emplaced. The arms and the back are held to the foot by two 90° curved aluminium bars 25 mm in diameter. This soldered structure is bolted onto the foot by a nut, soldered onto the 60mm tube. The chair leg is soldered to a 200x200x5 mm square base that is fixed to the ground by four M10 screws. To fix the chair in place the entire soldered base is sunk 10 cm below grade.

Seats

Nastra

Design: **Outsign**   Production: **Concept Urbain**

The Nastra set of seats comprises an armchair and a bench. The skeleton of both pieces is made of ductile cast iron with a shot-blasted finish. The armchair has a seat of laminated wood or a laser-cut galvanized steel. Both pieces come originally in dark gray; however, it is possible to order them in other colours. They are delivered fully assembled and have to be fixed by bolts or threaded rods.

Vesta

Design: **Jean-Luc Cortella**   Production: **Concept Urbain**

The Vesta bench has legs made of ductile cast iron with a shot-blasted finish. The seat and backrest are made of treated wood. The iron parts are available in any RAL colour. It is delivered fully assembled and has to be fixed by bolts or threaded rods.

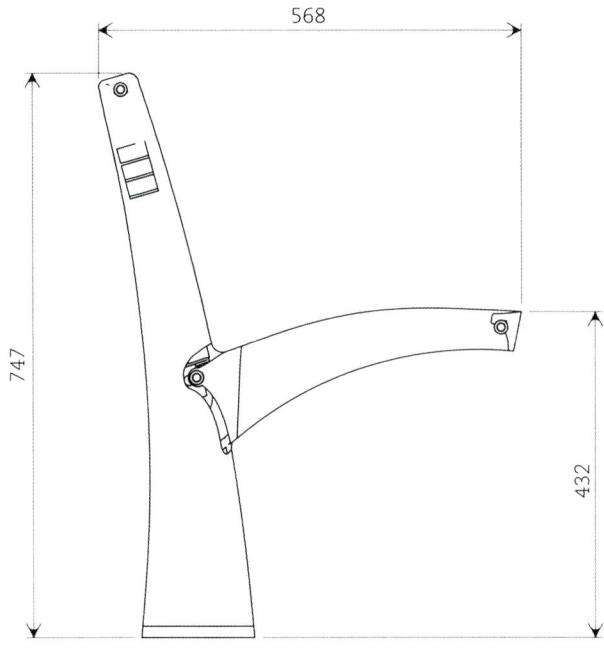

Loop

Design: **Lucas Galán Lubascher, José Luis Galán Peña & Roberto Fernández Castro**   Production: **DAE**

The body of Loop consists of 6 and 8 mm thick steel plate, folded and soldered, finished with corrosion-resistant powder-coated paint, over which powder-coated lead-free polyester steel-gray paint completes the necessary protection. Other colours can be provided on request. The seat is made of eight rods of tropical wood measuring 1852x40x25mm. The back has five rods 2000 mm in length. All the wooden parts are treated with tinted transparent oil that enhances the material's natural beauty and ensures maximum protection against the weather.

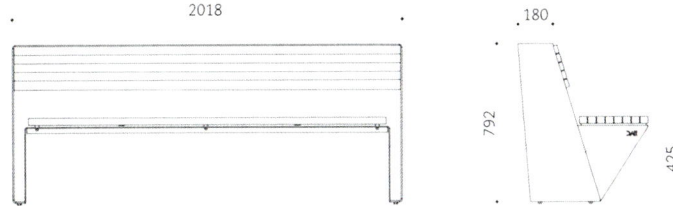

Pliegue

Design: **Jose María Churtichaga & Cayetana de la Quadra-Salcedo**   Production: **DAE**

The Pliegue chair is made of GG-20 cast iron, powder coated with polyester against corrosion. The item comes with a black wrought-iron-like finish but other colours and finishes can be supplied on request. The seat consists of eight rods of tropical wood, treated with translucent tinted oil to enhance the beauty of the wood and increase its resistance against the elements

Seats

Arco

Design: **Germán Rubio**   Production: **Durban Studio**

The shape of Arco breathes geometry, which is what makes it so striking. Ergonomically designed, this bench is appropriate for various types of users. In fact it consists of two items, a chair and a bench for the elderly, Arco-senior, which has been thought of specifically for those who make the most use of public benches. It is higher so that such users can sit down and get up more easily, and the armrest is wide for better support. The Arco bench is made of cast aluminium, lacquered, or cast iron finished with an anti-corrosive paint. The seat can be of treated tropical wood, autoclave treated northern pine, or composite board made of recycled plastics.

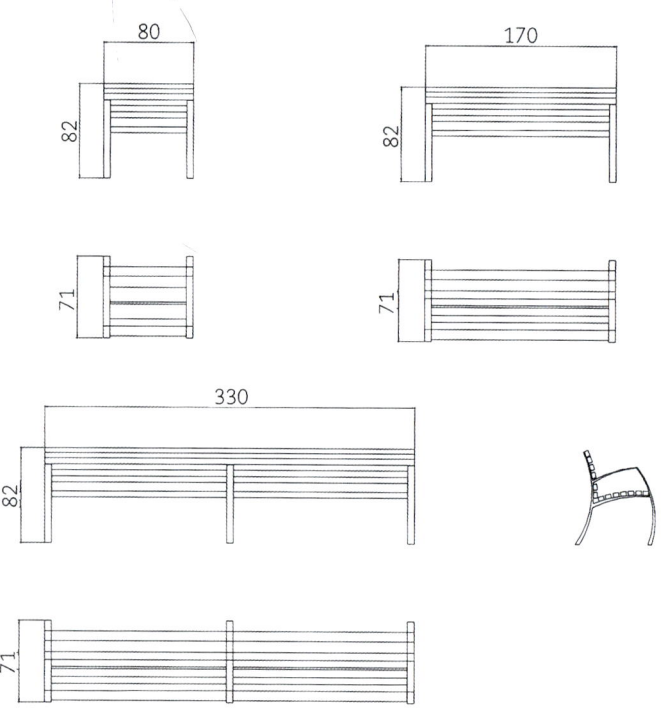

Hungaro

Design: **Equipo de Durban Studio**   Production: **Durban Studio**

Designing the Hungaro bench was preceded by a prolonged observation of the users and of the aging process of this type of structure. The design divides the surface and orders the space in a psychological way, offering more seating space than other benches of similar dimensions. The central trough prevents water from pooling on the top surface. The item is made of reinforced concrete, sandblasted, polished and water-proofed. Its weight makes any anchoring system superfluous.

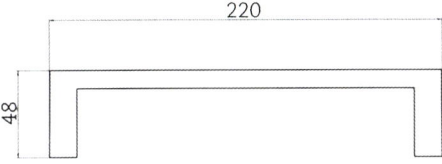

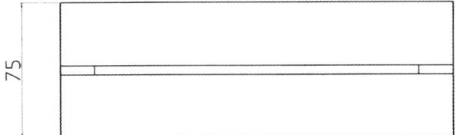

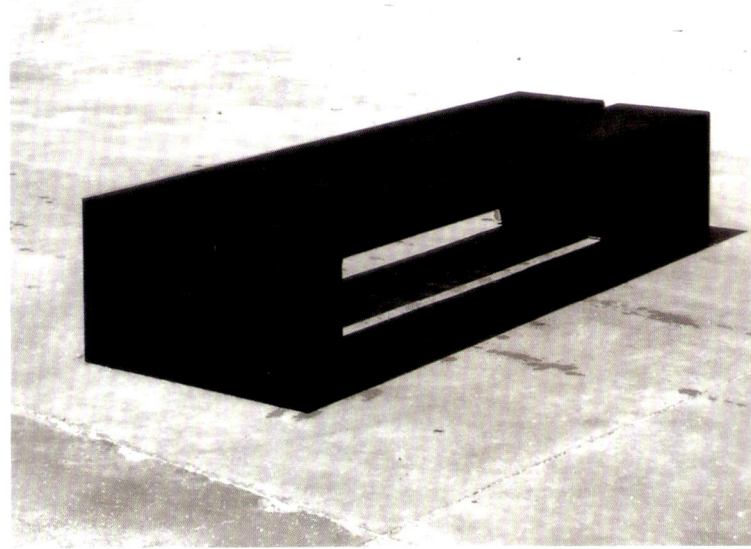

Lancer

Design: **Equipo de Durban Studio**   Production: **Durban Studio**

The forms of Lancer are defined by lines that are simple yet firm. These are ergonomic items that have a neutral character, enabling them to combine well in the spatial configuration of any public space. The piece is made of an innovative type of concrete that is reinforced with fibre-glass (GRC, glass reinforced concrete), making a thickness of only 5.5 cm possible without loosing its solidity. This guarantees the concretes durability and resistance even in highly saline environments.

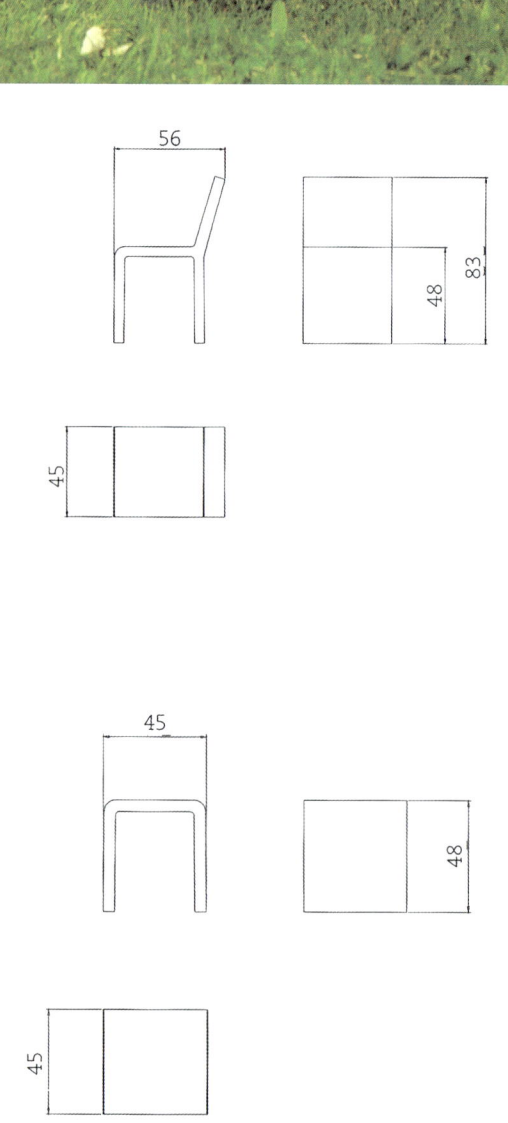

Seats 29

Armonia

Design: **Bernhard Winkler**   Production: **Euroform**

A comfortable bench consisting of a 15 mm round steel frame, hot-dip galvanised and powder coated in RAL colours, while the seating surface is made of expanded metal 3 mm thick.

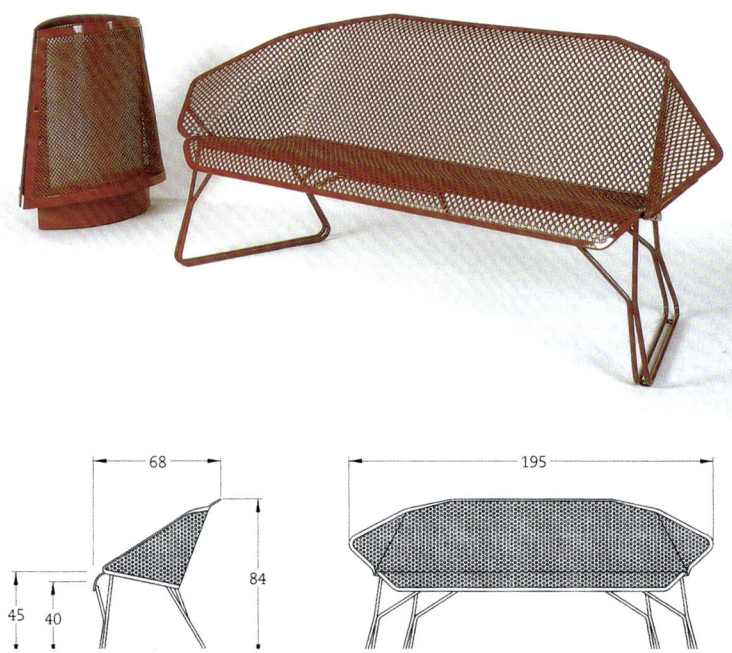

Fun Bank

Design: **Bernhard Winkler**   Production: **Euroform**

Fun Bank is an unconventional bench for kids and teenagers. With its unusual shape and resulting seating position, Fun Bank speaks the language of children. Fun is the perfect solution for any place where kids play, do sport or wait.

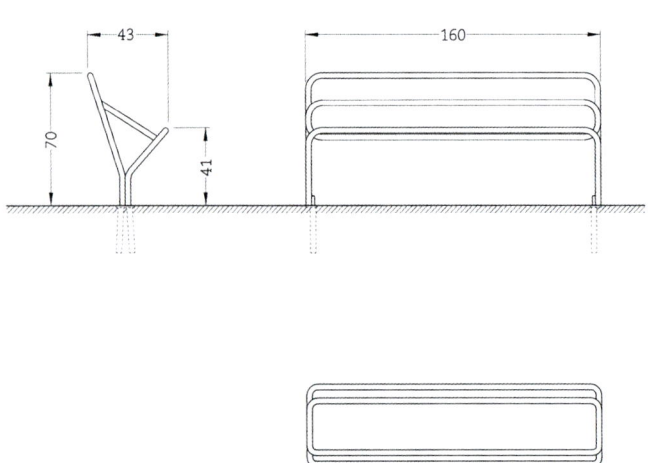

Buque

Design: **Diana Cabeza**   Production: **Estudio Cabeza**

Buque owes its name to its relation with a harbour and the context of shipping. In its design old traditions are respected but its image is modernized by using a line that links modern technology to the solidity of a historical location, the language of old ships of steel sheet and wood. The structure is of steel sheet with the soldering points showing. The finish is galvanized by immersion. The seat and the back are made of solid Lapacho or Quebracho wood, planed and waterproofed.

Seats   31

Pacú

Design: **Diana Cabeza**   Production: **Estudio Cabeza**   Development team: **D. Cabeza, A. Venturotti, D. Jarczak**

This is an individual armchair. Interesting compositions can be achieved by combining several of them, thereby stimulating social interaction in public spaces in the city. The seat is made of cast reinforced concrete with a tinted aggregate in the mix for permanent body colour.

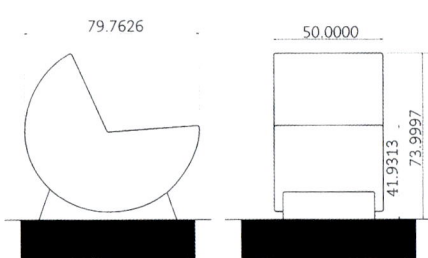

Patrimonial

Design: **Diana Cabeza**   Production: **Estudio Cabeza**   Development team: **D. Cabeza, A. Venturotti, D. Jarczak**

This urban element was designed specifically for the city of Buenos Aires. Both the version with a back and the version without it are made of cast reinforced concrete, with a tinted aggregate in the mix. The base is 1.20 m in length and permits the installation of individual armchairs or to place them in long rows, because they can lock onto each other. The possibility of facing them in opposite directions opens many opportunities of creating situations.

Picapiedras

Design: **Diana Cabeza**   Production: **Estudio Cabeza**

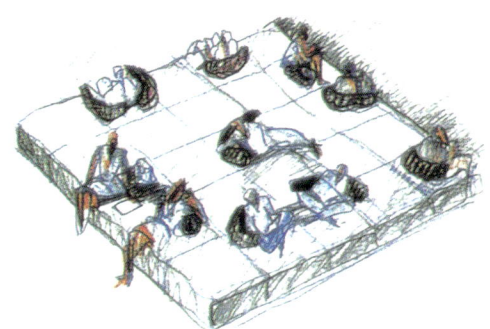

Picapiedras is a series of useful sculptures. They propose communal intercourse and are to be placed in strategic positions for meeting and exchanging information. They were designed for people to gather around them. The nature of the raw material is manifested almost fiercely and technology plays almost no role whatsoever. The stone is cut and the timber is carved in the most traditional ways, defining a topographical ensemble that becomes useful according to its form and the disposition that arises as the parts are combined. It is not a seating system, it is a usable topography that enhances freedom of use or possession and stimulates a sensual understanding of the raw materials. It enables a new relation to arise between the user and the object, creating an unexpected intimacy. The user can make contact with the materials, feeling the timber, its temperature, its ruggedness, its texture.

Seats

Linea

Design: **Thomas Winkler**   Production: **Euroform**

Linea stands out for its clean geometry and lines. The main architectural element – a rectangular steel rod, is repeated in all the products throughout the range. In this new line, designer Thomas Winkler has used a new approach to robust, durable materials and the well-known "euroform w" machining techniques, reintroducing them in a new, modern design. Linea products fit in perfectly with any setting due to a design that is both strong and classical. The entire range offers designers a great deal of creative freedom in customising the items for parks, town squares, and streets.

Sintesi
Design: **Bernhard Winkler**   Production: **Euroform**

Sintesi is clear, geometrical and extravagant – its aesthetic design lends an unmistakably modern accent to every situation. The frame consists of 70 mm gauge steel tube, hot-dip galvanised and powder coated in the colours RAL7016, RAL6005, and RAL9016. Other RAL colours are available upon request. The seating surface consists of 10.5 x 2.6 cm thick hard-wood boards, treated with a special varnish, and 1.5 mm wire mesh profiles. The piece can be fixed to the ground or remain freestanding.

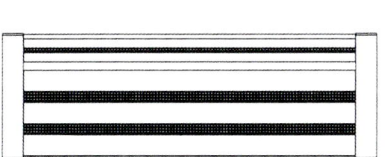
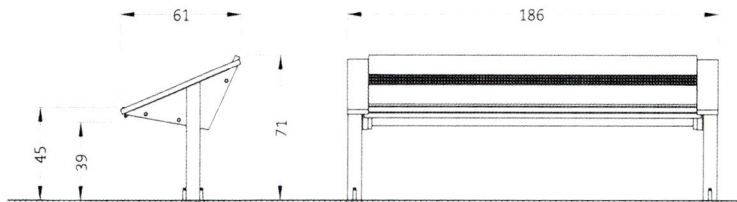

Alea
Design: **Josep Suriñach**   Production: **Fundició Dúctil Benito**

Alea is an ergonomic bench. The structure is made of steel and the seat is of tropical wood, measuring 150x45 mm. The finish coat is with Ferrus to protect the steel from corrosion and from saline fog. The wooden parts are treated with Lignus, a fungicide, insecticide and waterproofing agent that has no effect on the natural colour of the wood. The parts are fixed with M19 screws.

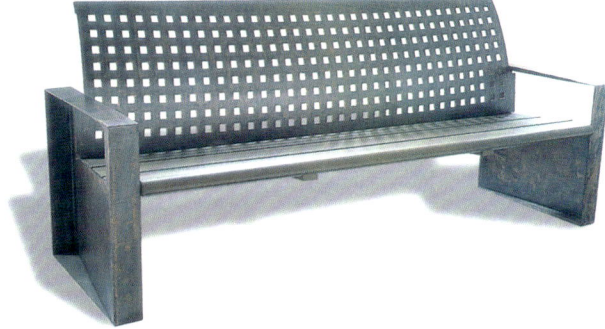

Seats

Metropol

Design: **Vilhelm Lauritzen** Production: **GH form**

The bench ends are cast iron, the legs are steel piping. The seat surface is delivered in stainless steel or wood. The bench is mounted on an embedded pin or secured straight onto the existing paving. The Metropol Bench comes as a single bench or a double bench, with or without a backrest. The double bench is available in a variant with armrests.

The Metropol Bench won the Brunel Awards 2001 international railway design competition. Comments from the jury: Very comfortable, elegant, rigorous, easy to maintain. Museum quality brought into public transport.

Victoria

Design: **Área de diseño Gitma** Production: **Gitma**

Victoria is a monolithic bench made of artificial stone. It was designed to be used at the end of a sidewalk or as a finishing or limiting feature of some sort. Its curved shape offers the user an amiable seating option which encourages conversations in groups.

Wing Pedestal

Design: **Rud Thygesen** Production: **GH form**

The Wing pedestal has a mahogany seat mounted on a steel structure. Wing was introduced at the Carpenters Fall Exhibition 2006, in Frederiksberg Garden.

Szekely
Design: **Martin Szekely**   Production: **JCDecaux**

Szekely is a bench with a capacity for six or seven people. It consists of a cast iron structure with two stylized central feet and a wooden seating surface on either side. This surface is made of wooden spars that are bolted onto the structure with screws protected with resin, creating a plane that seems to fold over the back of the seat. This determines a sober, symmetrical, minimalistic look, likely to enhance any natural or urban location.

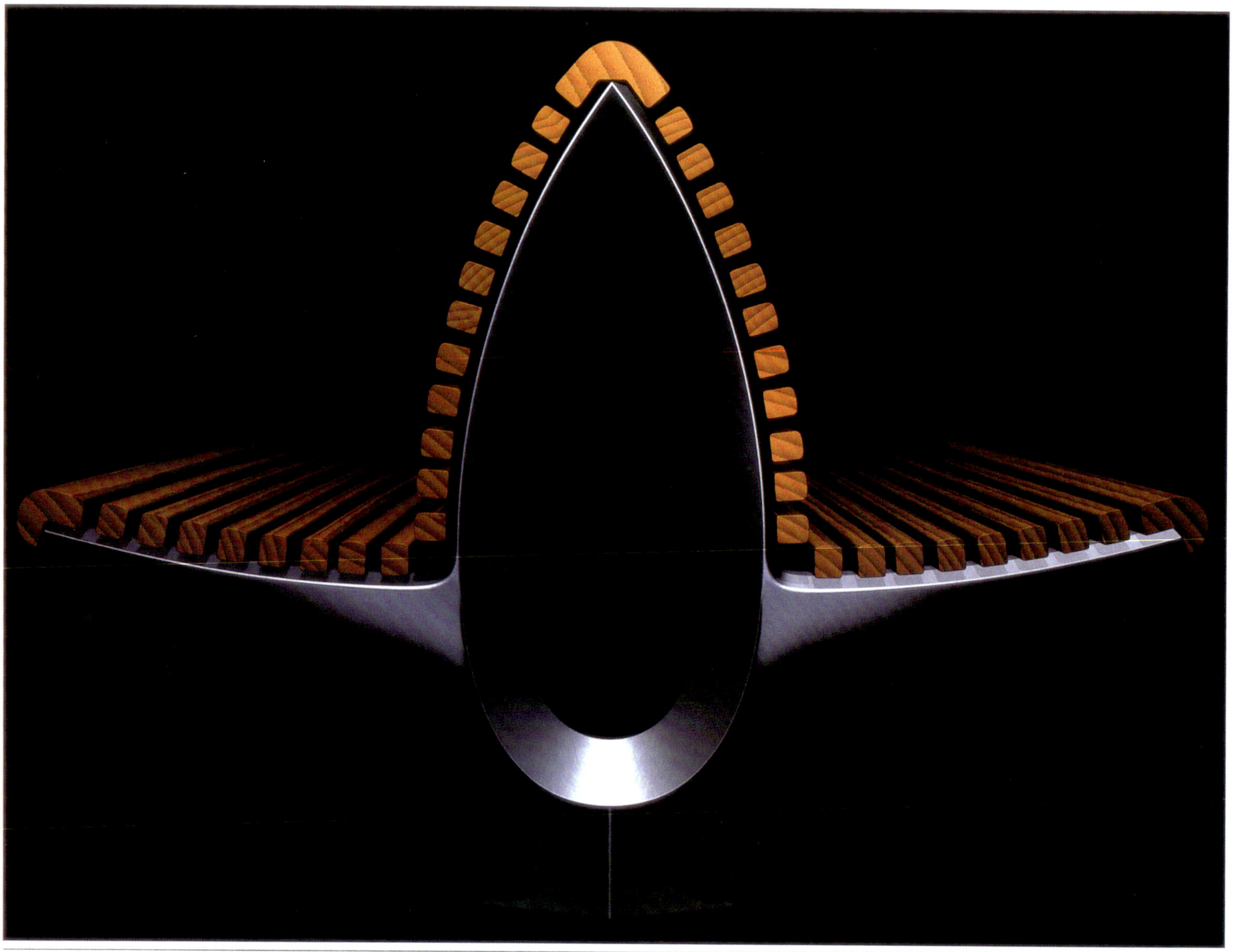

Pagoda
Design: **design-people**　Production: **Louis Poulsen**

The glowing street-chair emits light to both sides by means of its curved, white, inner surfaces. The duality of the chair makes it suitable for location in streets and squares, providing the area with low-level illumination and offering passers-by a place to rest. Pagoda is inspired in bollards but performs as an excellent part of outdoor urban furniture. The idea behind the product was to create an urban fixture with more than one function. Based on the Nimbus inground fixture, the Pagoda seat is a multipurpose lighting fixture suitable for all kinds of urban spaces. The design looks light but its construction is very sturdy as it consists of one single cast piece.

Daciano Da Costa

Design: **Daciano Da Costa**   Production: **Larus**

Daciano Da Costa directed this intervention, homage to Jorge Vieira (1926-1998), a sculptor in the Mediterranean tradition who gave his work a popular meaning while he resorted to an abstract expression of the human figure. His works mix well with the austere architecture of the city of Beja (Portugal). The benches make use of materials and solutions typical of Corten steel and of wastebaskets with an angle-iron foot, all of which is reminiscent of the three typical supports used for the sculptures of Jorge Vieira.

Rua

Design: **J.M.Carvalho de Araújo**   Production: **Larus**

The designers of Rua conceive of public space as a great sitting room, which should be comfortably furnished. It follows from this premise that urban furniture should be stable, fixed, oriented, vandalproof, repairable and changeable. It should not retain rainwater and should dry off quickly. Rua is a straightforward solution. A simple design could be the correct answer, with a neat and vigorous design susceptible of further adaptations. Moreover, it has a series of modular accessories to suit it to any space and type of configuration. Alternative thin films for the coating are being developed.

Degrau

Design: **Inês Lobo** Production: **Larus**

Degrau is a monolithic concrete design, the simple forms of which give it weight and solidity. Its shape makes it easily adaptable to sloped locations. From a distance, the sites where they have been installed look like an extension of a domestic living room into public space.

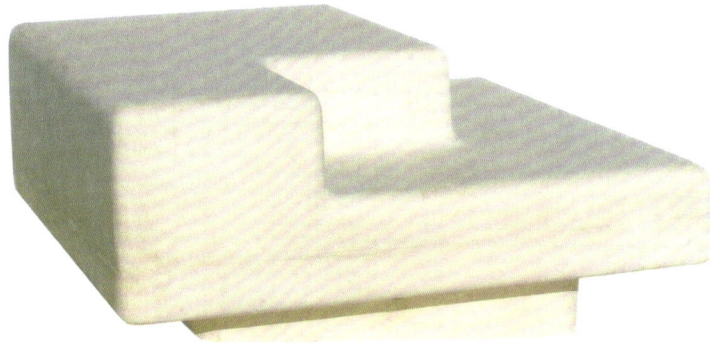

IP6

Design: **Jesús Irisarri - Piñeda**   Production: **Larus**

An ergonomic bench of roto-molded plastic for indoor and outdoor use. It features various possibilities of adaptation to different places. Its design includes drainage for any storm water that might gather. A specific anchoring system is provided to fix it onto the floor and it is illuminated inside. It can be supplied in various colours.

Coma

Design: **Joan Cinca** Production: **mago:urban**

This bench, made of sandblasted reinforced concrete, weighs 250 Kg. It attempts to respond to a wide range of necessities with one single object that can be used in many different configurations by simply altering the direction in which it is facing. Suitable for plazas, promenades or pedestrian areas, the basic configuration of the bench is ergonomic, but used in an inverted position it can become a limit or a bollard.

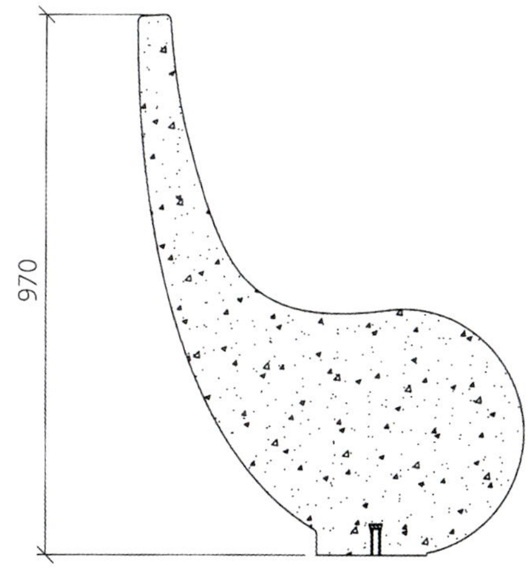

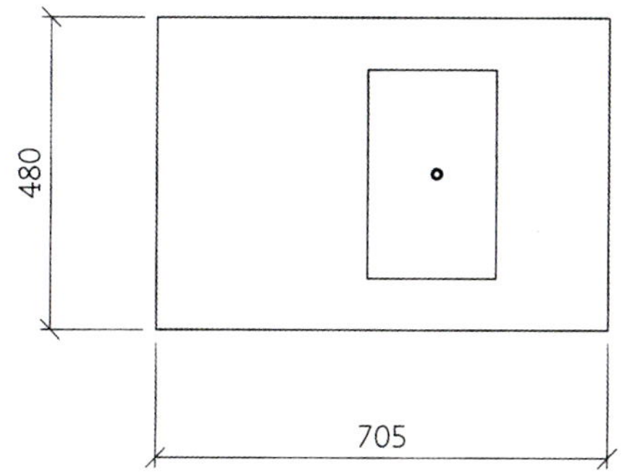

Kimba

Design: **Ton Riera**   Production: **mago:urban**

The Kimba Terra series comprises three different benches and two corner seats. With these five pieces it is possible to create a modular sequence of infinite possibilities. They will adapt to any requirement and can also be used as individual items. The three benches are differentiated by their outline, which is either straight, concave or convex. This multiplies the possible configurations with which to equip an urban space imaginatively. The two corner seats can be turned around either way, and thus adapted to all layouts and contours, however awkward this may be. When space permits, this characteristic can be implemented to alter the image of a location with a minimal investment.

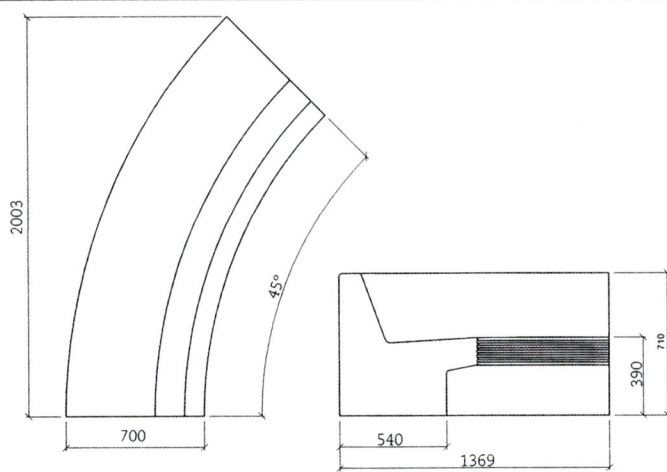

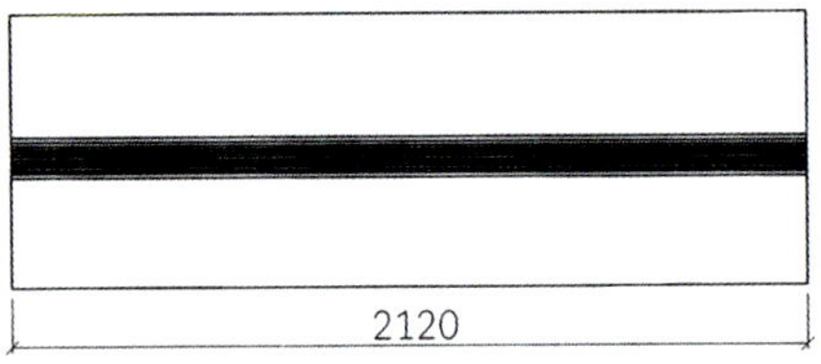

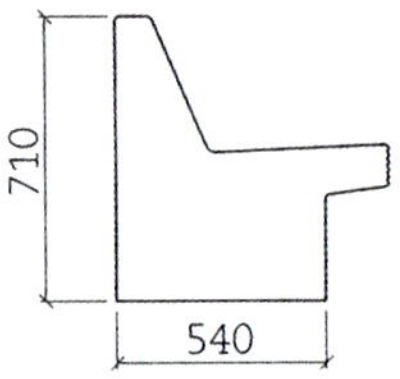

Terra

Design and Production: **mago:urban**

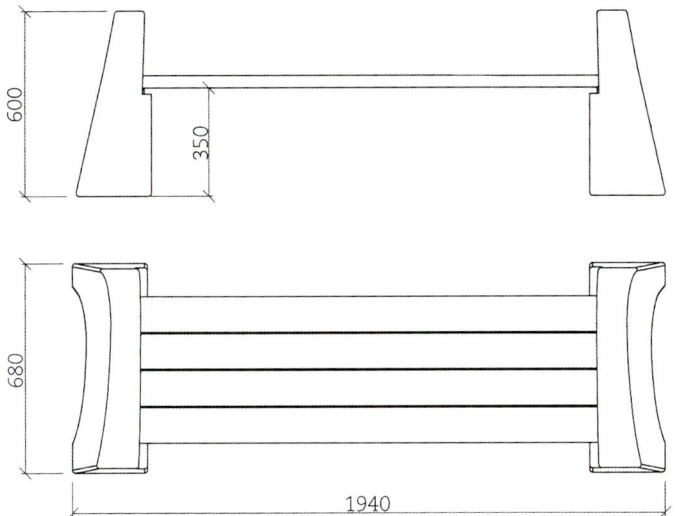

This functional and polyvalent bench is constructed with two supports of special concrete in the shape of an inverted "U". This design serves several purposes: anchoring system, armrest and support for the seat, which is made of crafted boards of tropical wood. The combination of wood and concrete allows this product to successfully fit into urban environments but also look its best in gardens, as it expresses the gentle warmth of wood and the tough rotundity of concrete.

Tube

Design: **Roger Albero**   Production: **mago:urban**

The Tube series consists of four bench models and an adjoining end-piece, which can be connected or used individually. Tube Silla and Tube Siesta are two ergonomic models that are perfectly suited to receive the human body in a state of repose. The body of all the models consists of a strip of reinforced concrete that surrounds a characteristic hole that defines the series and gives the piece lightness without endangering the sturdiness of the sandblasted concrete they are made of.

46   Seats

Seats 47

big bux

Design and production: **miramondo**

"big bux" is a multifunctional object: The design is based on the box tree, carefully pruned into the shape of a cube.
In keeping with its origins, "big bux" could and should be used as a decorative horticultural object (though it need be neither pruned nor watered). Thanks to its materials and structure, it is perfectly performant as a seat, but in conjunction with floor lighting, "big bux" becomes a lamp. Brightly illuminated from within, its leafy pattern casts exciting shadows onto the ground and neighbouring walls.

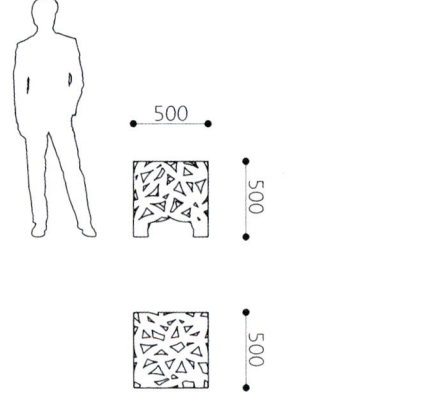

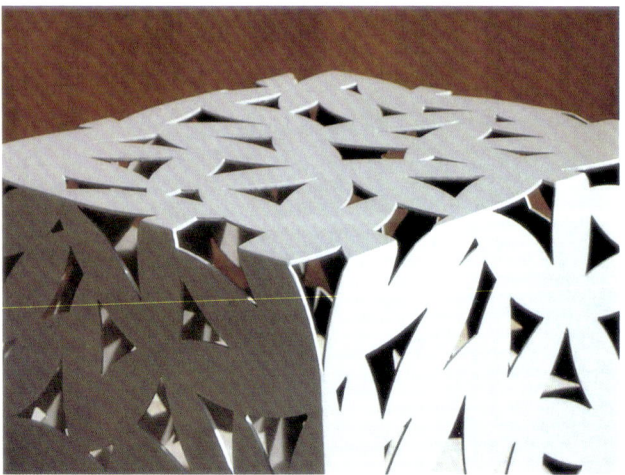

hop hop
Design and production: **miramondo**

"hop hop" is a broad range of products that can be used in a variety of ways. Thanks to their straightforward design, they can be mounted on walls and external steps etc. By means of additional brackets, "hop hop" can be cemented into the ground, remain a freestanding element or be mounted on a wall. By combining differing lengths of wood with the corresponding sidepieces, a variety of arrangements is possible. A remarkable feature is the formal resolution of the intersection between the seat and the backrest. By means of the vertical wooden slat on the frontal edge of the seat, "hop hop" merges with the surface beneath and conceals sharp or unsightly wall edges. The seats are simple to mount on walls and other horizontal surfaces using four screws and four special raw plugs with internal metric coarse thread. The wooden slats are connected to the sidepieces by means of a special screw fastener that can only be taken apart with a specific tool, making it vandal-proof.

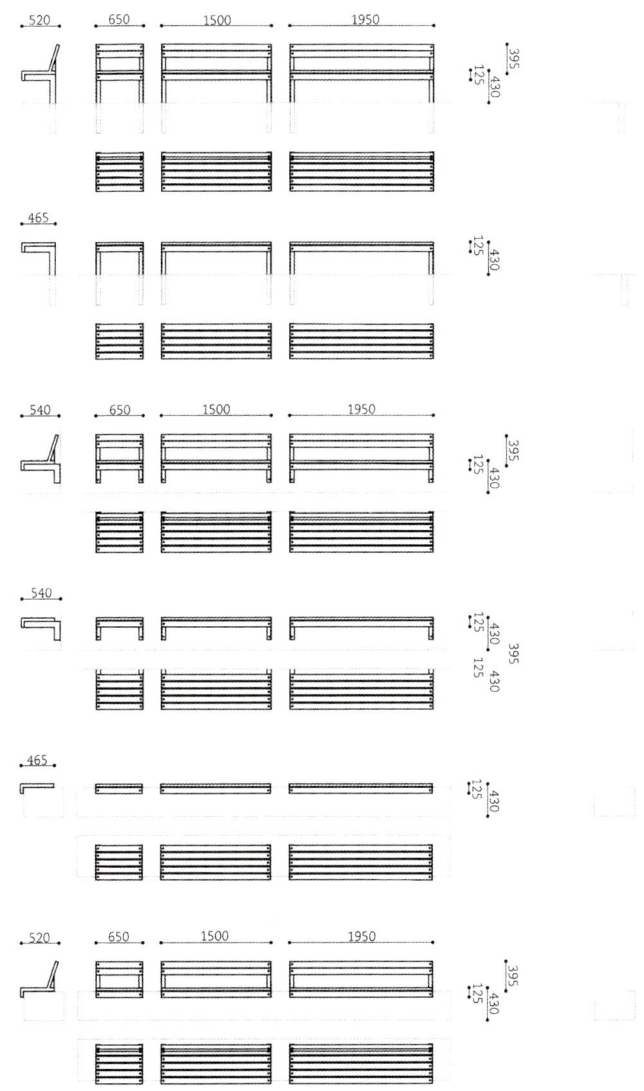

Seats

Brunea

Design: **David Karásek - Radek Hegmon**   Production: **mmcité**

A range of benches with a sophisticated modern design to bring elegance into every outdoor and indoor environment. Despite a light visual appearance, overall sturdiness is ensured. The new versions with either straight or curved arrangements of the transversal lamellas allow for variations and original (e) motional compositions. Brunea received the "Good design 2002" award. The hot-galvanised steel frame comes painted in a standard hue. The seat and backrest are either of solid wood lamellas or of perforated metal sheet, in both cases discreetly but firmly joined to the supporting frame. All four legs can be easily fixed to the ground.

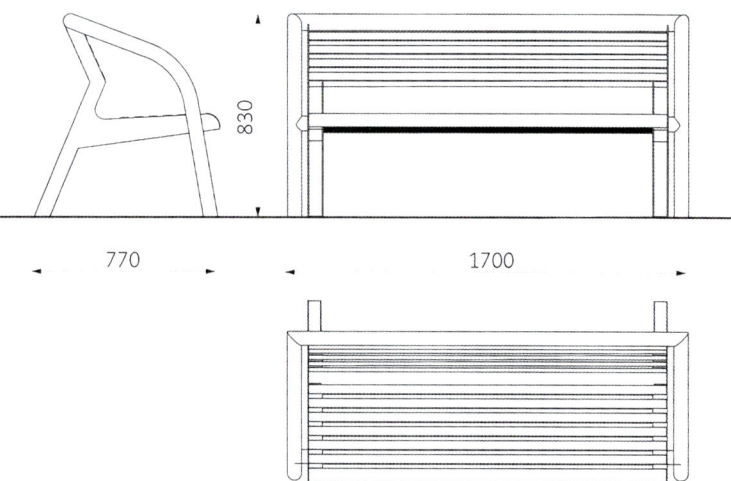

Laurede

Design: **Ruud van Eggelen**   Production: **mmcité**

This range offers robust outdoor furniture without neglecting a sophisticated design and optimum functionality. Its concise morphology will bring style to a wide range of different environments such as private gardens and parks. The structure of galvanized steel angle iron has been left without a surface finish, to weather naturally. The seat and backrest are made of solid wooden lamellas.

Katia

Design and production: **mmcité**

A range of comfortable benches with high backrests to make it easier to rise from the seat. Transversally placed ergonomic lamellas as well as side supports made of bent strip steel create an extraordinary effect. Optional models have a sectioned seat to prevent people from lying on the bench. The galvanized steel frame comes painted in standard shades. The seat and backrest consist of transversal solid wood lamellas which are discreetly but securely joined to the supporting frame. The steel frame is easily fixed to the ground.

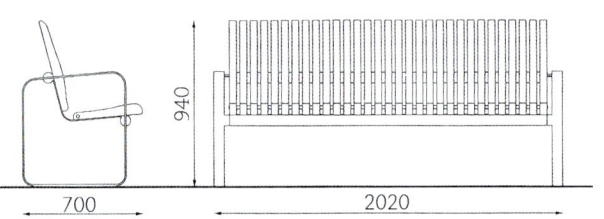

Elios

Design: **Tobia Repossi**   Production: **Modo**

This free-standing single seat in sandblasted concrete, treated to prevent deterioration, with an oval shape and a smooth surface, is usually supplied in a standard "grey" colour, but is also available in different colours (according to order). Elios does not require fixing to the ground.

Sitting-Around

Design: **Studio MAO - Emmeazero** Production: **Modo**

This item consists of a structure in stainless steel and slats in mahogany wood, specifically treated for outdoor conditions. The main feature of this unusual bench is its capacity to rotate, by means of a pivot placed at one of the two ends. An unconventional, variable and amusing seating arrangement!

Sol y Luna
Design: **Arch. Fausta Stella**   Producción: **Modo**

Sol y Luna consists of a structure made from steel and slats of solid mahogany specifically treated for outdoor conditions. Two supporting cylinders whose bases respectively represent the sun and the moon characterize this bench.

2197

Design and production: **Neri**

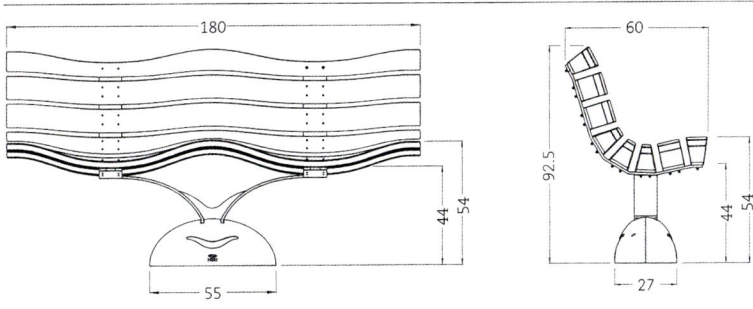

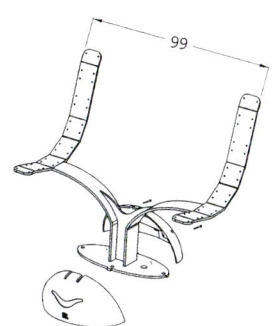

A fixed bench in UNI EN 1563 nodular cast iron and hot-galvanized steel, conforming to UNI EN ISO 1461 standards. The non-metal parts are of iroko wood (or equivalent). The bench consists of a hot-galvanized steel frame made of an elliptical flange, two cast iron half-shells decorated with bas-relief waves applied as covering for the flange and eight curved planks of laminated iroko wood.

Link

Design: **Nahtrang Design**   Production: **Escofet**

Designed by Nahtrang Design, the Link bench is an extremely light, urban-style seating element, but with a high formal content. Essentially recalling the classic folding stool, its structure is simple, attractive and balanced, with echoes of the East. A modest little gem whose proportions make it easy to integrate with the environment.

LINK's subtlety of form and smart combination of materials – steel and concrete or steel and bamboo wood – make it a balanced element in the urban landscape or for contract furnishing. The series includes the LINK bench, LINK stool and LINK wall seat models, all with two seat options, concrete or bamboo.

Seats

Longo
Design: **Manuel Ruisanchez**   Production: **Escofet**

The Longo series comprises six elements that make up four types of seats, a wastebasket and an ashtray. The bench is made of sandblasted and waterproofed reinforced concrete, in the colours beige and gray. It has a rectangular base that measures 400 × 100 × 45 cm. The seat of the bench, 2.8 m in length, is made of boards (section: 135 × 30 mm) of Bolondo wood. The bench has no back. The wood components are screwed onto three steel supports ready to be fixed onto the concrete bench.

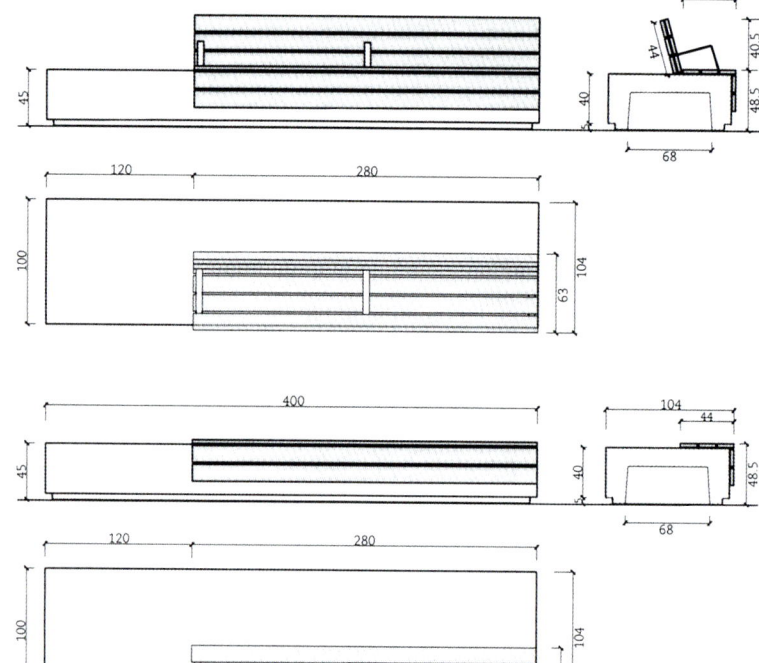

Pinxo

Design: **Coleman-Davis Pagan Architects**   Production: **Escofet**

Pinxo is a modular bench made of cast reinforced concrete. This item introduces a double purpose since it can be used to hang a hammock or as a conventional bench. It was Coleman Davis Pagan who first installed a version in dark red concrete as part of a project for public space on the island of Puerto Rico.

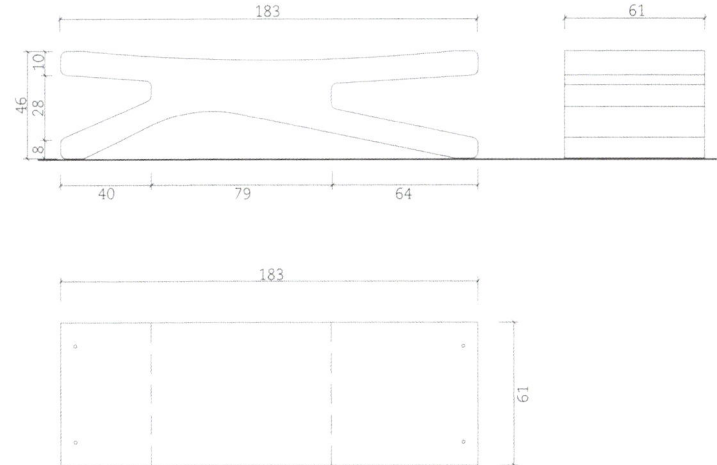

Seats   59

Twig

Design: **Alexander Lotersztain** Production: **Escofet**

Twig is a system of benches based on the concept of modularity, interactivity and connectivity. With these ideas as the starting point, the intention is to enhance the diversity and fluidity of the users. The design's versatility permits a series of meeting points to arise in between, generating new uses of public space. The material used is cast reinforced concrete and gently sandblasted, creating slightly rounded edges and smooth surfaces that branch out under the shade of a tree or on a university campus lawn.

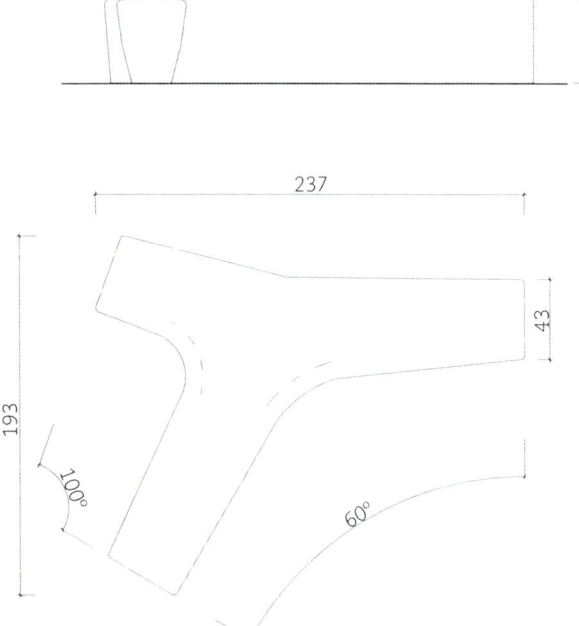

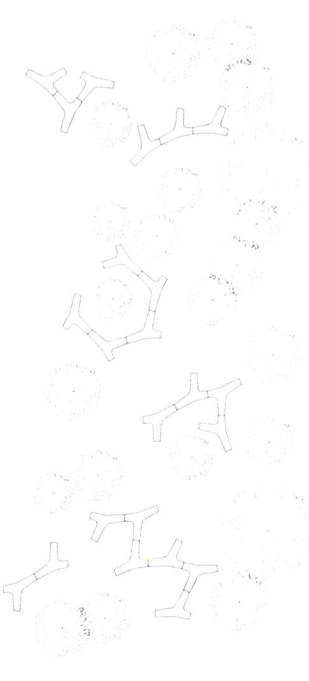

Seats 61

Yin-Yang

Design: **Francisco Javier Rodríquez**   Production: **Escofet**

The Yin-Yang bench is basically a reinforced concrete monolith with corrugated steel and a double finish. This brings out a diagonal axe of symmetry that generates a dividing line between the finishes, half sandblasted, half polished. The item's abstract formalization resolves the points it rests upon by means of two folds of the upper surfaces which, cut diagonally, reach down to the ground. The item is installed by means of two M-16 steel rods, 180 mm in length. Total weight: 475 Kg.

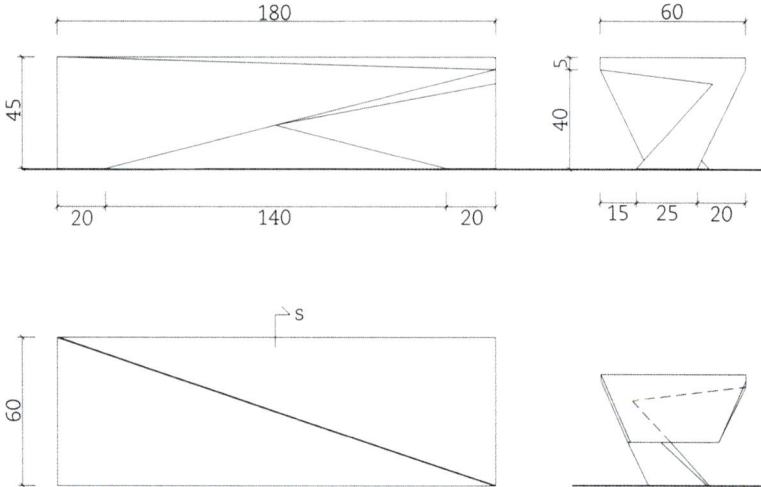

Alehop
Design and production: **ONN Outside**

The Alehop perch seat redefines the whole concept of seating for urban spaces. Its perch design makes it an ideal meeting point, allowing users to converse at the same height irrespective of whether they are standing or seated. This is not possible with a standard bench. This feature means the Alehop perch bench is particularly suitable for areas such as schools or universities, leisure areas, parks, beaches... It can also be used as a handrail to define areas. In terms of ergonomics, its height and the width of its seating surface ensure the comfort of the user-type it was conceived for.

It is produced in zinc plated steel, stainless steel AISI 316, or steel with an epoxy coating. The timber is Tali treated with teak oil.

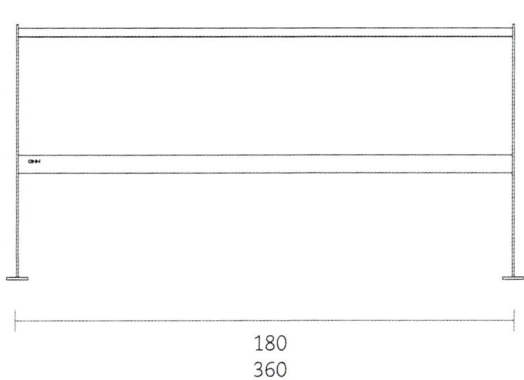
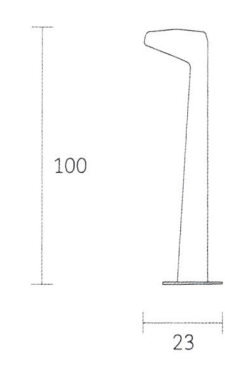

Seats 63

Tubbo

Design: **Paúl Basañez & Javier Machimbarrena**   Production: **Proiek Habita & Equipment**

The Tubbo collection is a solution for public space that can be equally well integrated in an indoor or an outdoor environment. This design represents an evolution of the "Romantic Bench" by means of a see-through structure of stainless steel tubes, whereby different locations are equally suitable. Its permeable appearance reduces its barrier performance and makes it doubly versatile. The series consists of a long bench, a short bench and a long stool without a back, plus a railing that completes the set, which should be understood as a system whose simplicity guarantees multiple possibilities.

Feris BK

Design: **Rehwaldt Landschaftsarchitekten**   Production: **Richtscheid Metallbau**

In the course of redesigning the market square of the city of Halle, a complete line of urban furniture was developed which includes bicycle racks, tree grates, trash receptacles as well as various benches and planters. The most prominent item is the "Feris BK" bench by Rehwaldt. The simple, timeless design of the bench smoothly integrates into the urban space. Powder coated steel cheeks combined with wooden battens give an example of how to offer a high-grade convenience and serve the needs of maintenance at the same time. A tiltable back is its main feature, making the bench a characteristic piece of urban furniture, accentuating the aspect of communication.

Mitrum

Design: **Rehwaldt Landschaftsarchitekten**   Production: **Kühn und Kirste**

Adjacent to a new research building and lecture hall in Dresden Technical University, two public squares were developed. The squares link the main axis of the building to the nearby streetscape. Sparingly arranged, but placed in a manner that is visually prominent, the site furnishing creates a sense of place and high-quality usability. Benches, bike racks and tree grates were especially designed for this site. The seats and benches have battens of wooden cubes. The warm coloured scumble leaves the material's texture visible in contrast to the concrete stand. Ruled by the grid, the design refers to the architecture around it and communicates the scientific purpose of the building.

Board

Design and production: **Runge**

For those who love to socialise, Board consists of top-quality timber formed into a half-oval shape using traditional handcraft techniques.

Interchangeable, weatherproof varnished or active-ventilation painted, this bench offers pure recreation whether it is used for sitting or lying on, with unlimited possibilities and variations.

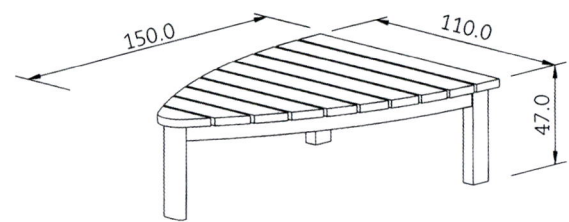

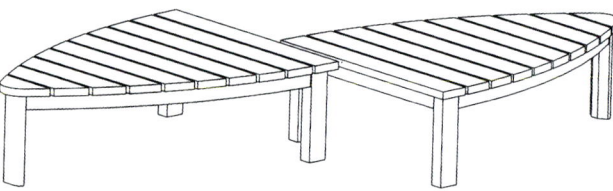

108

Design: **Enric Batlle - Joan Roig** Production: **Santa & Cole**

This modular urban bench in rhomboid shapes is made of mass coloured concrete. The concept arose from an angle, designed to shape limits, contain and form alignments, with or without a backrest. Its geometric characteristics only permit curved alignments with large radii. Its simplicity and discretion make it fit into any urban space without prominence. The standard line is mass coloured in greyish tones, stripped finish, freestanding on the ground. Different finishes and colours can be available depending on the quantities ordered. It can also be manufactured with recyclable aggregates. It can be delivered in two combinable measures, 48 cm and 150 cm. The line is complemented by a piece of the same height and shape as the backrest, aimed to contain the ground. The designers recommend a minimum distance of 3 cm between the benches.

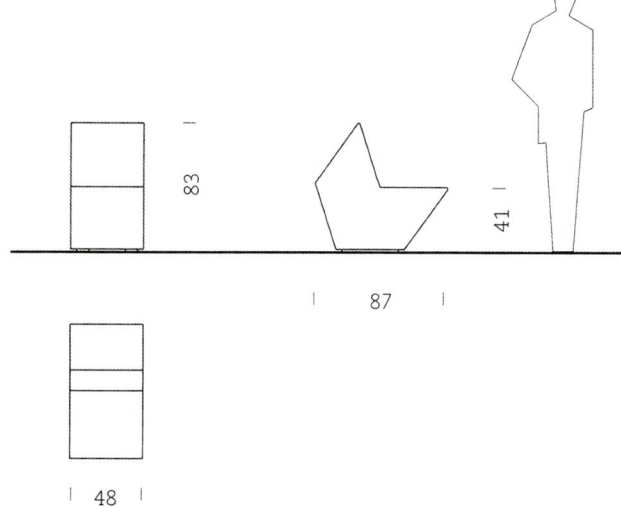

Seats 71

NeoRomántico

Design: **Miguel Milá** Production: **Santa & Cole**

This is an adaptation of what is now the classic NeoRomántico bench, with arms and a lighter structure. It is designed to complement urban or private areas needing an even more comfortable bench than its predecessor. Made of cast aluminium and wood, the bench arose from the need to update the NeoRomántico bench while adapting to the existing NeoRomántico market, thereby redesigning this romantic seat. The wide and rapid acceptance of the article in urban areas is the result of both its appearance and its comfort. Unlike its predecessor, its structure is more geometric and lighter. It is a unit designed for use in private areas and public or domestic use. It is remarkable because of its lightness and comfort. All the benches come with arms. It is solid, ergonomic, versatile and clean.

Bancal

Design: **Julià Espinàs - Olga Tarrasó**   Production: **Santa & Cole**

This is a robust and simple bench designed for public meeting areas. It is conceived as a modular and versatile unit to provide long alignments of seating surface, with or without backrests. Its formal clarity predominates over its rhetoric value. The Bancal bench is especially designed for public use and has centuries of tradition. It is modular, solid, has a sober appearance, is designed for open spaces and is made to withstand adverse conditions. It is a linear bench for public areas and may come with or without a backrest. It consists of a steel structure and wooden boards, both common and long-lasting materials. It is designed to respond to urban plans for large benches with different needs and orientations. Its elemental but thorough design permits the backrest to be placed in two positions in the same alignment or simply omitted altogether.

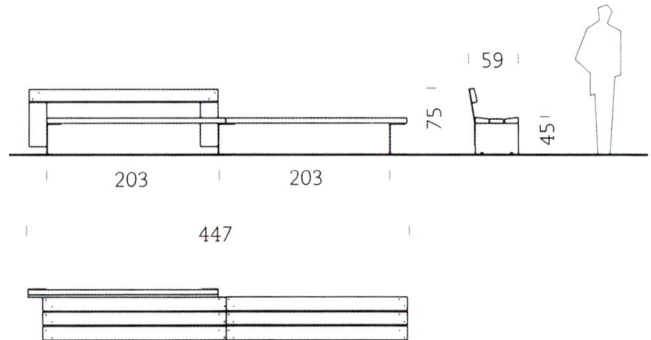

Adelaide
Design and production: **Street and Park Furniture**

Street and Park Furniture are committed to the view that better streets and civic spaces are a vital part of community well-being, and endeavour to be perceived as a principle supplier of street furniture for these applications. They aim to provide specifiers and customers with the convenience of a single supply source for most urban projects incorporating street furniture, backed by specialised expert support. Materials: Polished cast aluminium and bronze, steel, and recycled Australian hardwood, oiled.

Agora
Design and production: **Street and Park Furniture**

Street and Park Furniture were given a brief to create a seat that was durable, virtually maintenance free, could have variable installation details and would be provided in different lengths. Having ascertained that stainless steel would be the best material, the next task was to find a mesh, wire or punch pattern to use as the infill between the profiles. The designers found an architectural wedge wire mesh originally used as shower grates or in outdoor drains. The transit seat can be installed on the ground, surface mounted or wall mounted. It is also available as a 2, 3, 4, 5 or 6 seater.

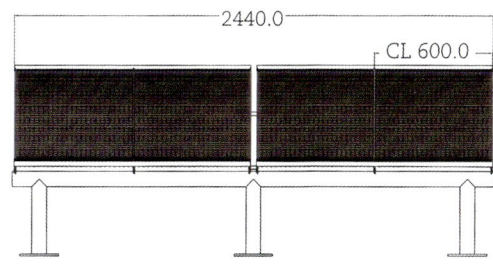
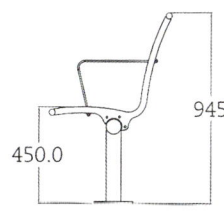

Seats

Shoreline

Design and production: **Street and Park Furniture**

The full range of Shoreline furniture also comprises deck and cube style benches, a picnic setting, fencing, a custom-made wave-style cast bollard and a spiral bike rack. The furniture was designed to be very simple and streamlined. It was originally designed for a new foreshore development in South Australia. Since then it has been installed on neighbouring foreshores as well as being contracted for other coastal developments. It has taken over the coastline!

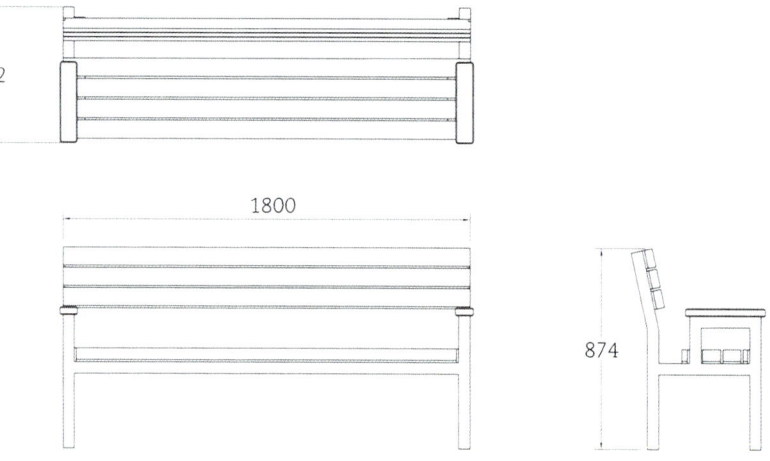

Horse Shoe
Design and production: **Streetlife**

The Horse Shoe bench is minimal and distinctive. It consists of metal legs with rounded FSC Cumaru slats, a very elegant and hardwearing combination of materials that can withstand the ravages of time. The dark-red colour of the wood differs slightly in tint from slat to slat. The elevated seat with the round form of the horseshoe creates a surprisingly pleasant yet active seat. It adapts well to architectural plans, both in and out of doors. It is also available in a broader version that enables double-sided use. Leveling the bench is achieved by adjustable stainless-steel legs that end in studs to be simply embedded in the ground.

Longlife
Design and production: **Streetlife**

The Longlife benches are the longest standard benches on the market, available in 2 lengths, as an 8 or 12-seater. They offer simple natural seating and enhance the linear dynamics of urban landscape. The FSC hardwood slats are laminated and finger jointed with a high-tec bonding system. The FSC hardwood is lively, durable, and will never rot. The slats are mounted in the patented Streetlock® comb system. The stainless steel mounting materials are theft-proof. The Streetlock® comb makes it easy to replace a slat. The slatted seat comes pre-assembled. The legs and brackets are galvanized.

Rough&Ready
Design and production: **Streetlife**

All Rough & Ready products are sturdy and robust. This unique program comprises: Straight benches, Linked benches, Curve benches, Circular benches, Topseats, R&R Bicycle Parking racks and Lineparking, Bollards, Tree tubs and Tree isles with R&R sitting rims. The All-Black beams of recycled agriculture plastics are another available option. The R&R Curve system, the R&R Circle benches and also the R&R topseats are designed for outstanding and natural spaces.

In conjunction, they form a coherent program for exceptional urban locations. They are fixed to the thermo-galvanized legs by means of the Streetlock® comb system, which is stainless steel and theft-proof. The optional backrests can be fixed in various positions.

Tauranga

Design: **Ted Smyth** Production: **Woodform**

Designed by Ted Smyth for Tauranga, New Zealand, this bench is made of solid Kwila wood bent using the Woodform technology. Woodform Design has pioneered research into technology to develop systems for the bending and shaping of solid timber and MDF (Medium Density Fibreboard).

Divano

Design and production: **VelopA**

With its round lines and streamlined design, the Divano is an exceptional bench. The gently curving seat and backrest are made of durable wood and provide optimal comfort. The length gives a sense of allure as well as providing plenty of seating space.

Seats 81

Ara

Design: **Marta Ferraz, Paula Cabrera & T&D Cabanes**   Production: **T&D Cabanes**

Ara is a bench that combines formal equilibrium, ergonomic principles and top quality materials. The boards on the seat can be of tropical wood or of polythene in different colours. The seat rests upon a curved structure with angular steel supports. The result is a wide, comfortable and inviting piece of urban furniture. The benches can be combined together to form different configurations. Alone or in groups, standing in a row or randomly placed, they create a variety of dynamic compositions.

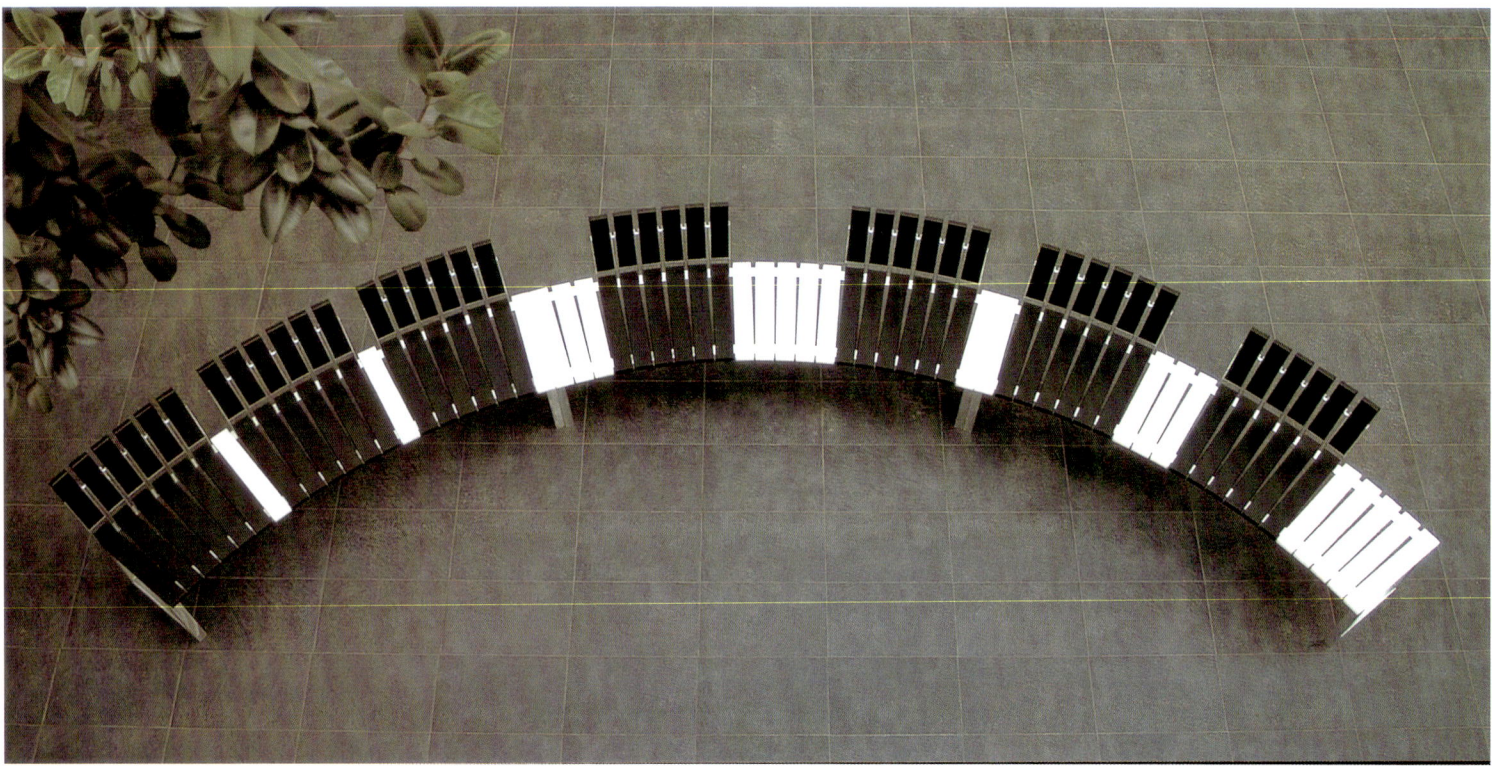

Bluemoon

Design: **Vicente Soto**   Production: **T&D Cabanes**

The Bluemoon bench features a conceptual design. Its ergonomic structure is a composition of purist forms, executed in an interplay of steel and tropical wood boards. The backpiece consists of one single board. It has been designed to enhance large, open air surroundings and gardens, although it performs to equal advantage next to highly technical architecture where metal prevails.

Sombra

Design: **Vicente Soto**   Production: **T&D Cabanes**

The Sombra benches combine compositional balance and formal lightness. They consist of a single support of carbon steel that performs as the main anchoring element with the ground and is the joining point of the thick wooden boards of tropical wood that make up the seat and the back.

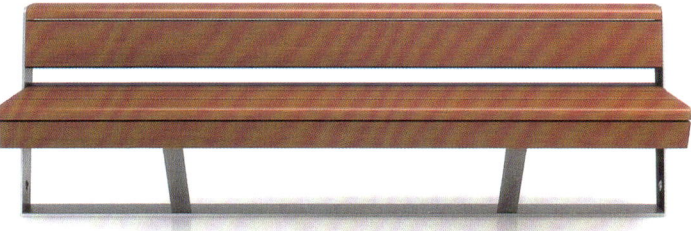

Seats   83

Celesta

Design: **Javier Feduchi, Alfredo Lozano y Pablo Moreno**   Production: **T&D Cabanes**

The main characteristic of the Celesta bench is the rationality and the solidity of its form. It consists of a monolithic body of saw-cut crystal granite, and a structure of stainless carbon steel, fixed to the ground by means of a chemical adhesive.

Rehué

Design: **Diana Cabeza**   Production: **Estudio Cabeza**

Rehué is a set of seats and tables of two different sizes, which can be placed in multiple configurations, for outdoor or indoor use, for private or public space. The pieces are constructed of cast concrete using tinted aggregate. The standard choice of colours available is earth and black.

Cado

Design: **Max Wehberg**   Production: **Westeifelwerke gGmbh**

The Cado seating series is based on a simple trapezoidal foot that sustains a curved sitting surface. Different configurations can be created based upon this element: with or without a backrest and with a solid or a lighter appearance.

So-ffa

Design: **BCQ Arquitectos** Production: **Escofet**

Soffa is a bench that doubles as a piece of minimalist sculpture. Its dynamic form is made of reinforced concrete, coloured white or black. Its solidity makes it unnecessary to anchor it to the ground in any way.

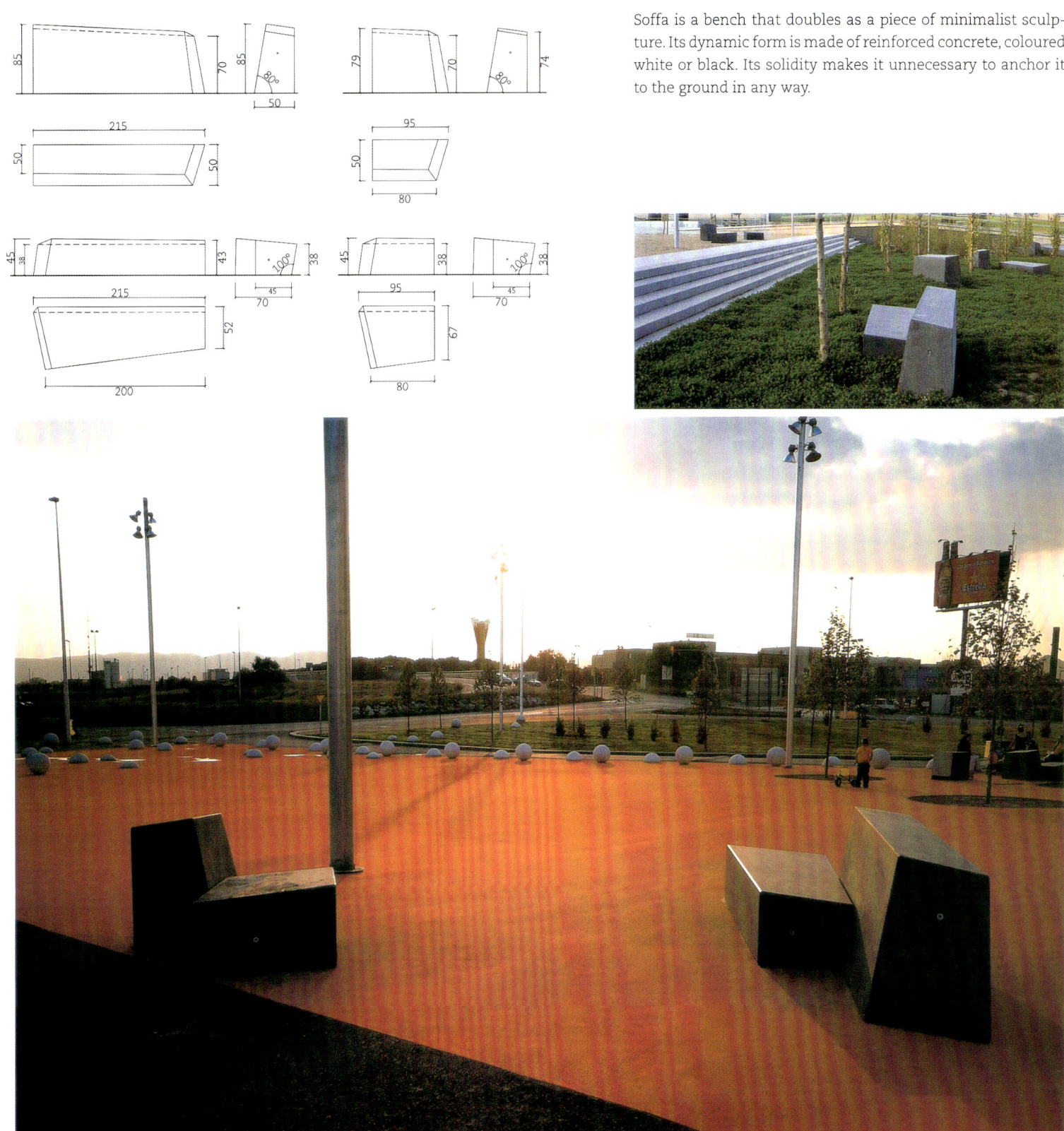

Moiré

Design: **Alessandro Piaser**   Production: **Modo**

When a pattern of closely set parallel lines is superimposed on another such pattern aligned slightly differently, the interference pattern causes a visual effect similar to moiré silk. The Moiré bench is inspired in this effect. It consists of a seat and a backrest made of parallel stainless steel rods, Ø 10 mm and Ø 16 mm thick, fixed onto supports of the same material, laser-cut out of stainless steel plate in a shape that defines the outline of the bench. This is stabilized underneath by Ø 60 mm diameter steel tube that crosses the full length of the construction and rests on a base of Vicenza stone. An alternative anchoring system allows it to be fixed to a wall, or 100 x 100 mm legs that can be fixed to a concrete edge. The bench is available in 2000 mm modules, with a straight or a curved outline that comes in two versions, concave or convex. The object's dimensions are 1,020 mm in height and 680 mm in width.

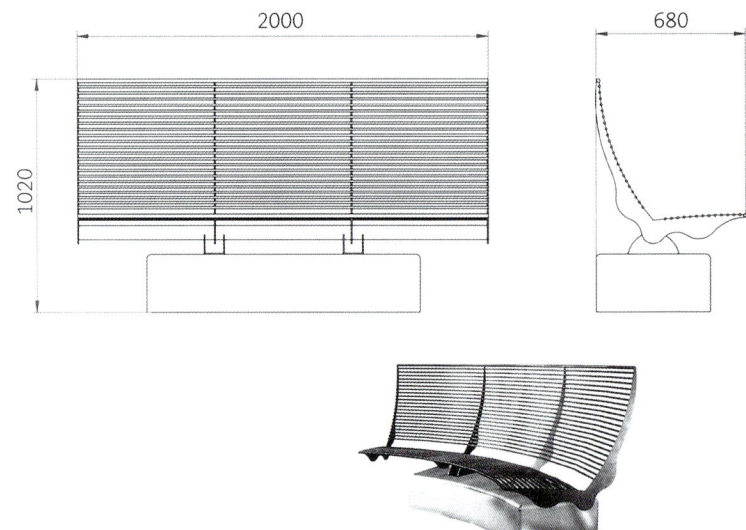

Sumo

Design: **Enric Pericas + Josep Muxart**   Production: **DAE**

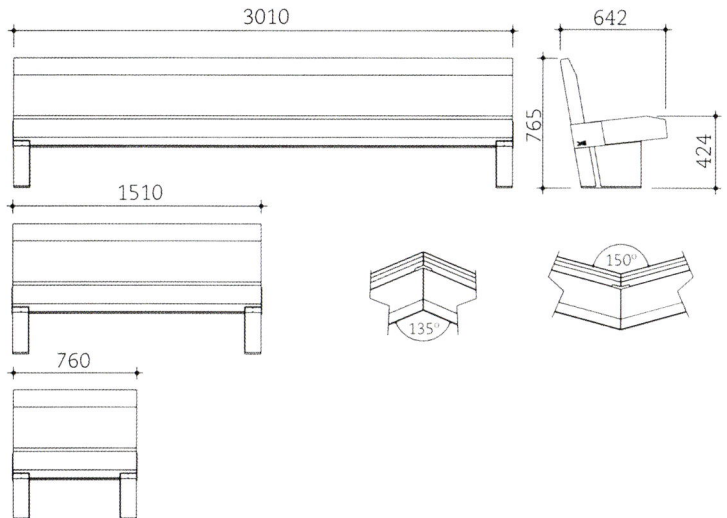

This bench consists of a seating surface and a backrest made of composite pinewood planks, high pressure impregnated, sawn and glued, plus three hot-dip galvanized steel legs. The seat and the backrest are treated with a base coat of deep-penetrating tinted base varnish and two layers of semi-gloss, transparent, lightly tinted alkyd resin finish varnish. The cross-sectional dimensions of the seat are 442 x 140 mm, and the backrest measures 535 x 80 mm. The item is supplied in three standard lengths: 3000, 1750 y 750 mm. The legs are fixed to the ground by Ø10mm expansion bolts.

Dom

Design: **Sergio Fernández**   Production: **Tecam BCN**

The Dom bench comes in two forms, a simple backless bench and a complete bench with backrest and armrests. The backless Dom is a simple urban bench made of five pieces, which include the seat of four 30 mm (1.18 inches) thick planks and two legs of steel tubing, with optional armrests. The planks can be of Elondo wood or Flemish pine, autoclave treated and with two coats of a waterborne varnish. The legs are galvanized steel, fired with micro textured silver. The bench is mounted with stainless steel screws.

Radium

Design: **David Karásek, Radek Hegmon**   Production: **mmcité**

This new product line derives its unique contemporary characteristics from the aesthetics of bent sheet steel. The optimal rigidity and excellent overall resistance of this steel bench is achieved thanks to the ingenious intersection of its walls. The version with a wooden seat presents a different visual aspect. The seat surfaces of both options have a rounded longitudinal slot to avoid lounging. The structure is made of galvanized steel sheet, painted in standard shades; the seat consists of metal sheet or solid wooden boards (optionally, round steel bars can be used). It is provided with an easy anchoring system for fixing to the concrete or pavement. This product line also includes new models of litterbins and bollards. A small stool extends the possible applications. The series was distinguished with the "Excellent Design 2005" award.

Seats 93

The Swiss Benches
Design: **Alfredo Häberli**   Production: **Bd**

The Swiss Benches were created for Bd as a homage to the famous Banco Catalano. The outcome is a number of models of public benches that offer that extra function so typical of the designs of Alfredo Häberli. The Poet is a bench at which one can eat or write, with a table to rest on. The Banker is somewhat harder, The Philosopher is visually tranquil. Other suggestive options are The Loner, for lonely hearts, and The Couple. The toughness required of a public bench, which must resist corrosion and vandalism, need not be incompatible with aesthetic quality. The Swiss Benches are available in two optional finishes, hot-dip galvanized or painted with polyester resin. Both display all their beauty either indoors or in the open. (Translator's note: In several languages, the word bank indicates both a financial institution and an elongated seat).

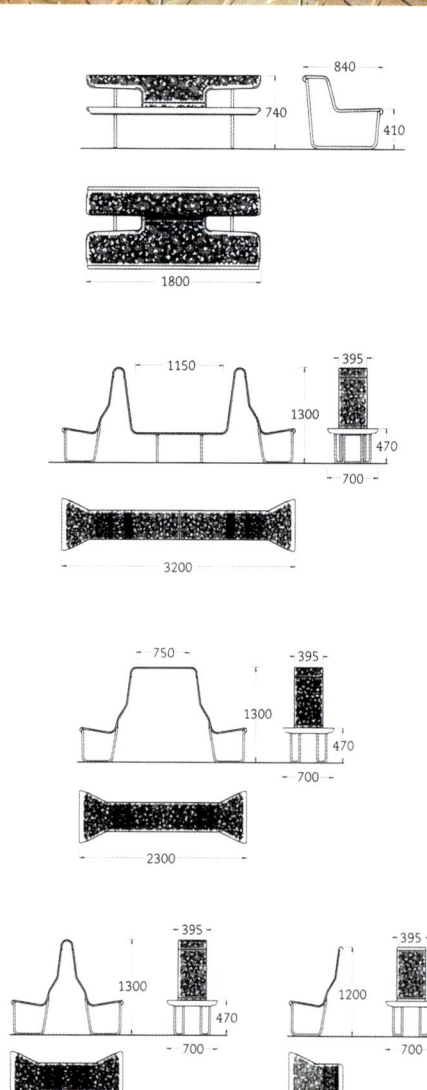

94   Seats

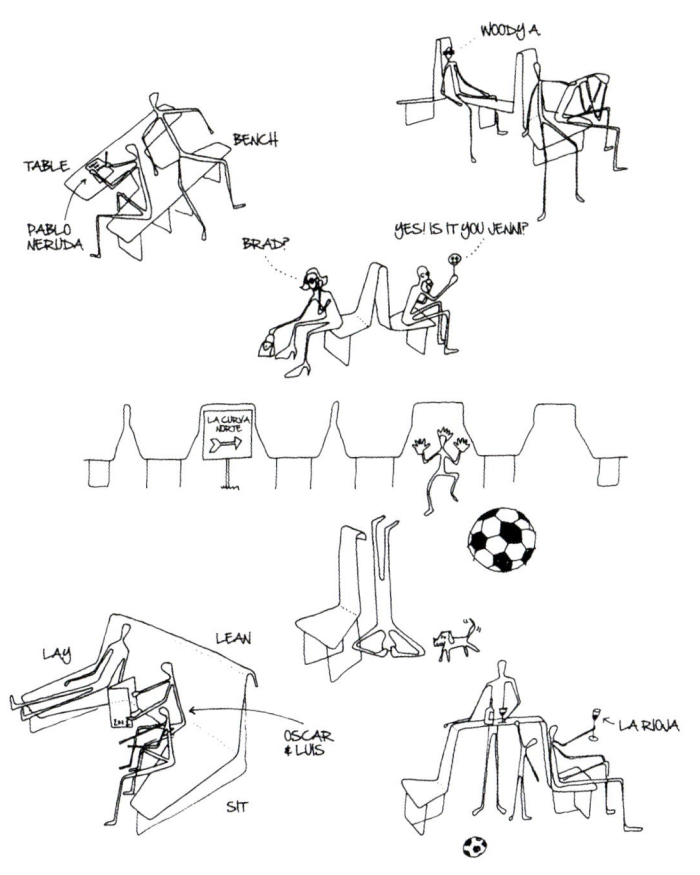

Racional 2
Design: **Franc Fernández**   Production: **Tecam BCN**

Reducing the idea of a seat to the fundamental basics, this bench consists of a horizontal and a vertical component and two pressure-impregnated (autoclave treated) pinewood planks. These two main functional parts are supported by a hot-dip galvanized steel structure that provides the legs and the armrests. Only the two front legs and the armrests play an important visual role, as the two back legs are discreetly placed near the middle, where they prevent the seat from bending. The legs are connected and braced by the rest of the structure, which is concealed under the seat and behind the backrest. Racional 2 is 278 cm long. Taking the resistance of the materials to the limit and optimizing their performance was the motivation for this design brief.

Botanic

Design: **Michelle Herbut** Production: **Street and Park Furniture**

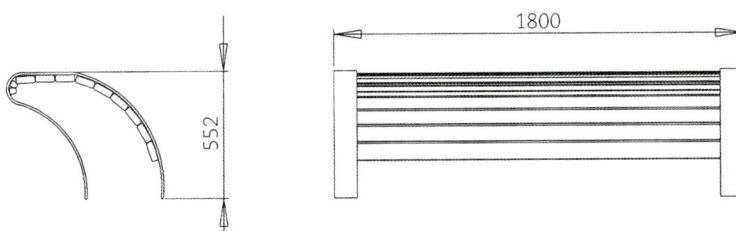

1800
552

Michelle Herbut, who graduated as a Bachelor of Industrial Design in 2005, designed the Botanic Bench for a final year University project to suit a specific site in South Australia. Inspired by the growth and organic nature of the plants at the Botanical Gardens, she wanted to simulate the bench 'growing' out of the ground and took advantage of the resulting curve to add another possibility to the ways you can use the bench, by using it as the much sought backrest that nature so often omits. Street and Park Furniture, with whom Michelle currently works, were approached to help with production and manufacturing issues. With their specialist experience, the enterprise provides urban projects with the convenience of a single supply source of street furniture, backed by specialist experience and support. The Botanic Bench consists simply of a series of hardwood boards bolted onto the two curved end pieces made of aluminium.

Cuc

Design: **Foreign Office Architects (FOA) - Farshid Moussavi & Alejandro Zaera Polo**   Production: **mago: urban**

The Cuc (worm) bench was designed to fit into and complement the paving system of the Auditoria of Barcelona Park, one of the emblematic spaces of the Forum of Cultures Area. This paving system consists of units that, placed one after the other and rotating on their own axes, allow it to adapt to any sort of terrain, however irregular. The bench could not conform to straight lines or Cartesian geometry, as the paving was designed to follow the sinuous profile of the dunes in the park. This was the idea behind the Cuc bench.

Made of concrete, the Cuc seat consists of a truncated cone that tapers away from its round seat towards the ground. It has a groove down one of its sides, which locks into the next unit, and so on. The surface of the seat curves up to provide a small, radial backrest. The Cuc bench is capable of multiple configurations simply by rotating each of the units that compose it. It is possible to construct straight or curved structures as further units are added to the chain, resulting in an increasingly flexible shape as the chain gets longer. The geometry can be re-altered even after being fitted. The Cuc units are made in three colours: white, granite grey and ochre.

Accesible

Design: **Diana Cabeza**   Collaborators: **L. Heine, M. Wolfson**   Production: **Estudio Cabeza**

This urban bench is ideal for resting or reading in an open public space. The different lengths of backrest and seat, along with their constructive independence, allow alternative uses such as a backrest for those sitting on the ground, or lean-to for someone standing or even a protective enclosure for prams or wheelchairs. The backrest ends with a horizontal surface at the top, for use as a support for reading material or to lay out the picnic.

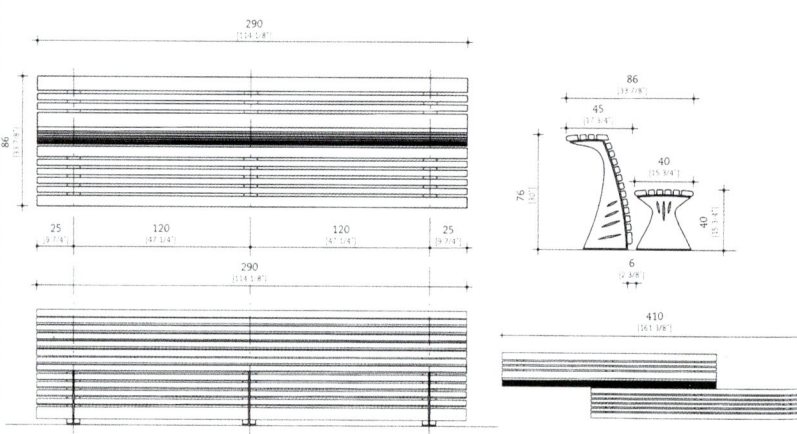

Rambla

Design: **Diana Cabeza**   Production: **Estudio Cabeza**

This system of public benches offers great freedom. Organised in rows, the benches constitute a flexible support with log-like backrests. Randomly placed they are an invitation to multiple and varied uses.

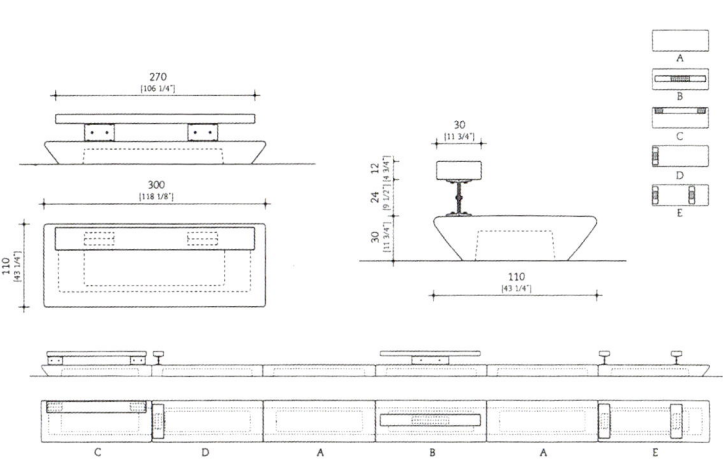

Seats

Godot

Design: **díez+díez diseño**   Production: **Escofet**

The design of the Godot bench is based on the idea of creating an element of street furniture that could be integrated into Samuel Beckett's stage-set for the play from which it has taken its name. The bench program consists of three prismatic-shaped units (single, double and triple), respectively measuring 92, 152 and 212 cm (37.33, 59.73 and 83.31 inches). These elements are made of concrete and all have a semi-circular hollow at one end. Assembled contiguously, they are designed to leave an empty space inside the module to be used as a planter. Godot can be installed around trees, forming a kind of pedestal that transforms and highlights the plant, while inviting passers-by to enjoy its shade.

Boomerang

Design: **Andreu Arriola / Carme Fiol** Production: **Escofet**

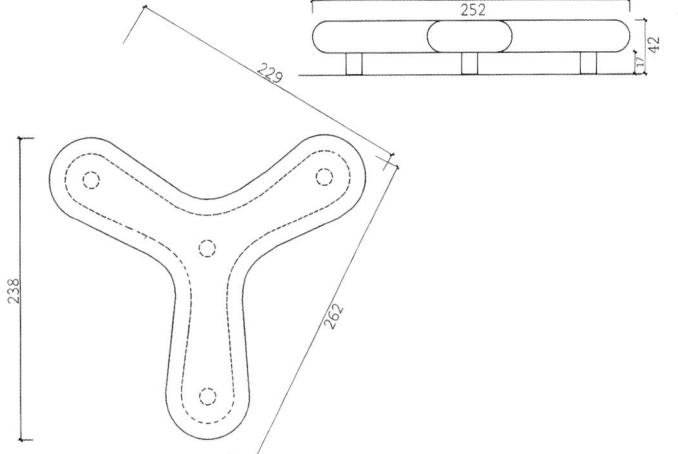

Boomerang is one of a set of seats and tables made of stone. Their friendly shape invites the passers-by to rest, read or deposit what they are carrying, constituting an innovative reminder of ancestral nature and the world of fossils and seashells. These items were designed to form part of natural parks or green areas, to furnish plazas, promenades or neutral public areas in the city with an object that is rational while offering a welcoming natural counterpoint. They are outdoor sofas that can be installed in large indoor spaces such as sport pavilions, exhibition spaces, major facilities, markets and such. Boomerang introduces itself as a playful, conversational item that symbolically embodies communication and movement.

Wastebaskets

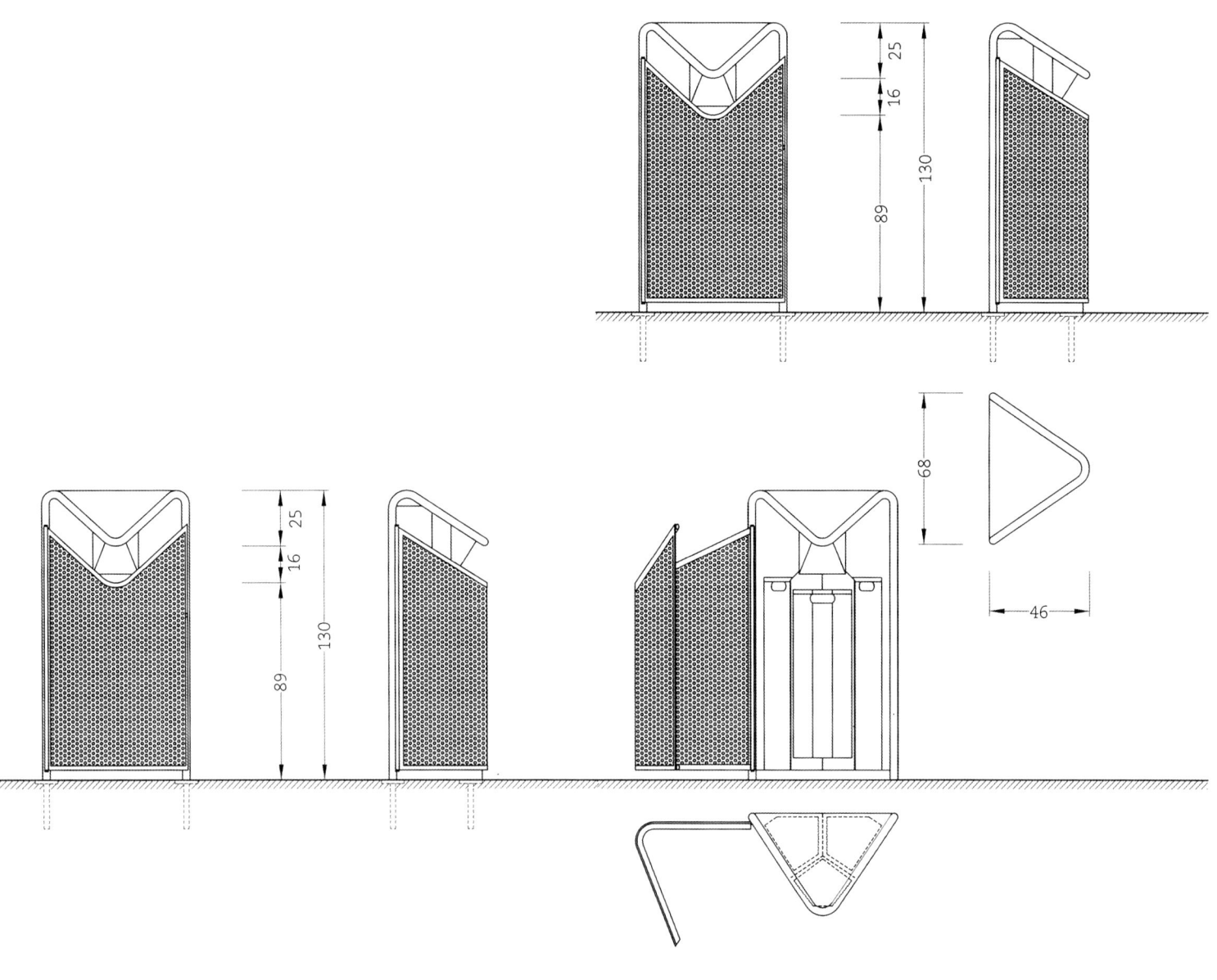

Sloper

Design: **Luis Tabuenca** Production: **ONN Outside**

Sloper is a complete range of street furniture designed by Luis Tabuenca for the Tabuenca Saralegui and Associates studio and developed and manufactured by ONN Outside. Biomechanical research of the human body has established that an angle of 76° is the most suitable for the backrests on street seating. This angle has become a geometrical ruling criterion for a range of products that meet all the requirements of exterior environments. Sloper provides the solution to any urban project using the same aesthetic criteria of practicality and top quality materials. In addition, the flexibility of the lights, which allow for use and optimization of whatever light source is preferred, makes Sloper compatible with numerous street lighting projects.

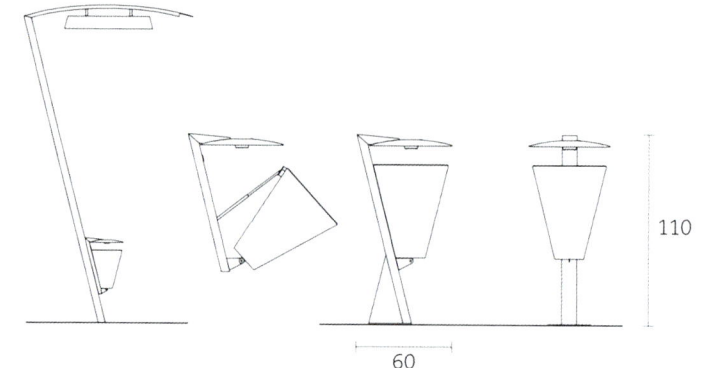

7kale

Design and production: **ONN Outside**

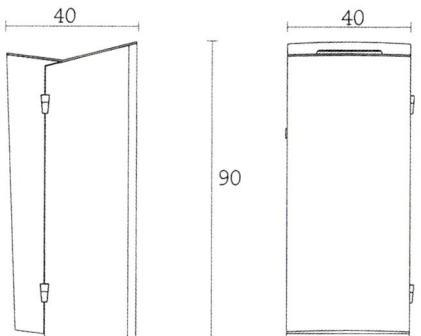

The 7 kale litterbin gives the impression of two asymmetrical bodies joining brusquely. This product has a distinctive, up-to-date feel, which can be appreciated from different angles. For better ergonomics in the cleaning process, this litterbin has been designed with a door at the front with a triangular lock making for easy access. The liner has two plastic handles for easy removal. It is suitable for use both outdoors and indoors and is available in two versions: wall-mounted (40 litre) or freestanding (70 litre), either fixed to the ground or on a pedestal with a steel counterweight hidden in the interior. A removable ashtray is included and the bin may be personalised with adhesive stickers or laser engraved with coats of arms or logos. The 7kale litterbin was selected in the 2003 ADI-FAD awards.

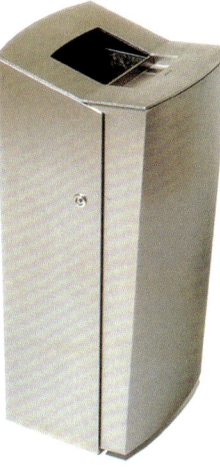

Wastebaskets

Laurel & Hardy

Design: **Gonzalo Milà / Miguel Milà**   Production: **Escofet**

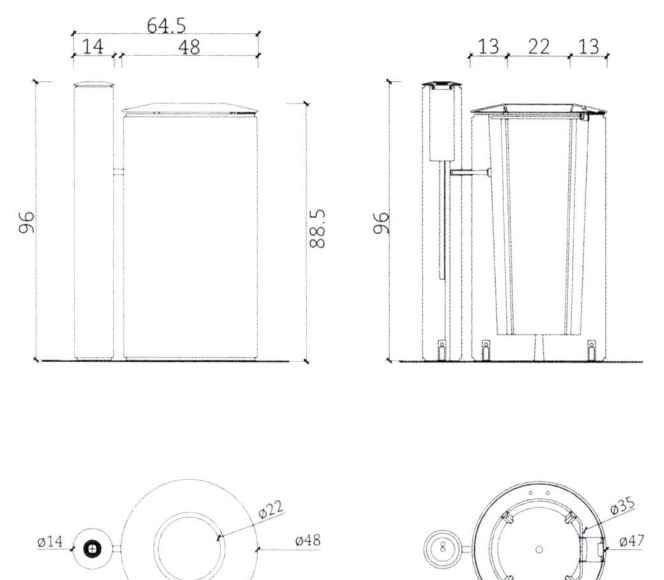

The Laurel and Hardy waste disposal unit has a twofold function as ashtray and wastebasket, tasks that it can perform autonomously or together. It consists of artificial-stone cylinders finished off around the top with a ring of cast aluminium. These items can be installed in heavy duty locations, indoors or outside. The wastebasket has a lock to prevent vandalism and it should be used with a plastic garbage bag. The container of the ashtray can be taken out for it to be emptied into the wastebasket without disassembling it altogether. The two cylinders are fixed together by a threaded bolt and are likewise bolted to the pavement. It comes in two available finishes, gray concrete and smoke-black aluminium details or beige concrete with the aluminium details painted to look like Corten steel.

Net

Design: **Diego Fortunato**   Production: **Escofet**

Net is based on the idea of a line of urban furniture that is gently rounded and recalls no particular historical period. It is an attempt to address the sensual quality of cast concrete, enabling it to integrate well into any environment. Athough it comes in standard tones of black and white, it can be supplied on request in the full range of colours produced by Escofet. The surface texture has been gently sandblasted. Each item has a capacity of 40 litres and weighs 155 kg. It requires anchoring to the pavement for stability. The plastic container-bag is fixed by means of an articulated stainless steel ring. The compatible garbage-bag size is 575 x 1000mm.

Ecology

Design: **Bernhard Winkler**   Production: **Euroform**

The Ecology wastebasket is a functional design conceived for places that require different wastebaskets for different kinds of waste. Its design allows three different types of waste to be collected in individual compartments.

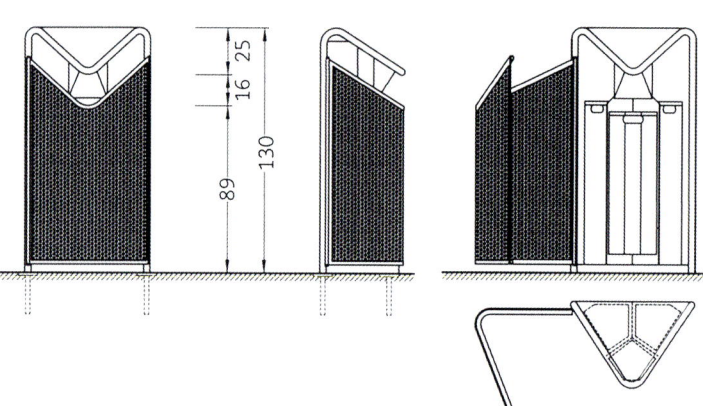

Quattro

Design: **Bernhard Winkler**   Production: **Euroform**

The elegant, well designed Quattro litterbins always ensure a clean environment in their many designs, with or without a cover, fixed to the ground or mounted on a wall. It is available in two versions with or without the cover, which provides for an ashtray. It has a capacity of 50 litres.

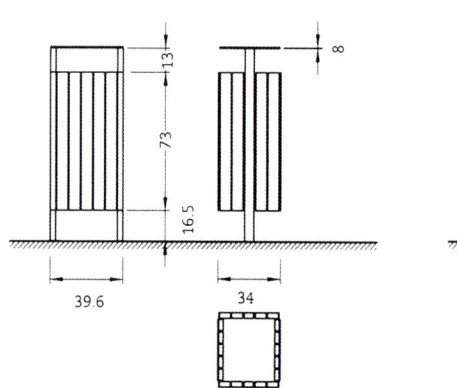
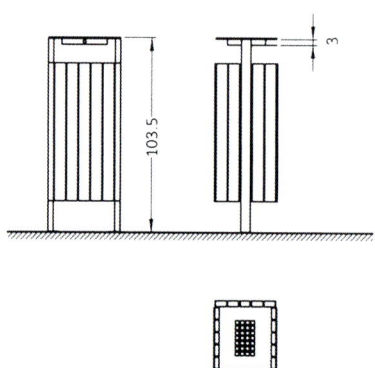

Imawa

Design: **Urbanica**   Production: **Concept Urbain**

The Imawa wastebasket is made of galvanized steel finished with RAL lacquer and has a stainless steel grill decoration. It has a sack holder or interior aluminium liner with a handle and a volume of 60 litres. Its front opening has a locking system with a special key and stainless steel lock.

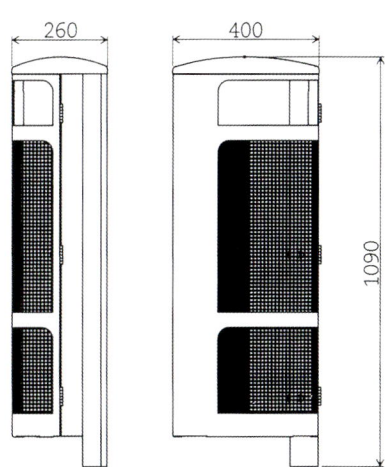

Nastra

Design: **Outsign**   Production: **Concept Urbain**

Nastra has a body of cast aluminium, finished with RAL lacquer in the colour of choice. Its 85 litre interior is lined with aluminium. It has a locking system with triangular key. The stainless steel lock has a returning spring.

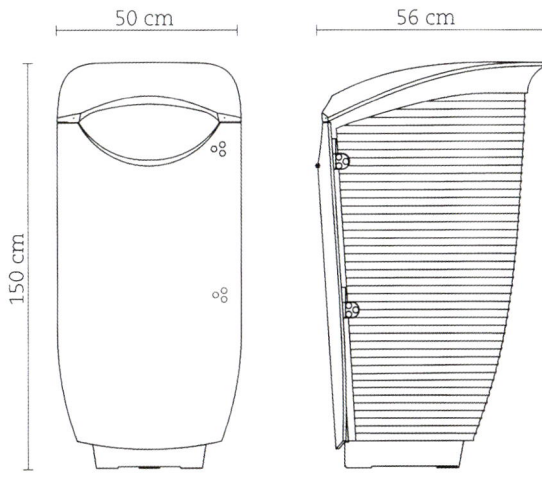

Banquina

Design: **Estudio Hampton / Rivoira y asociados**   Production: **Estudio Cabeza**

Trash, one of the lines designed by Estudio Cabeza, includes the Banquina waste bin. The item is made of 2mm thick perforated steel plate with a ½" thick rim guard. The support posts are of solid, untreated, Quebracho wood. The object is finished with a coat of thermo-setting powder paint.

377

Design and production: **mago:urban**

The 377 waste bin is a piece of urban furniture designed and produced by mago:urban, characterized by its cylindrical shape and topped by a polyester lid that is also the mouth of the bin. The opening is fairly small to avoid the users depositing large garbage bags that would make it necessary to empty the bin with excessive frequency.

Sacharoff

Design: **Enric Pericas, Maria Luisa Aguado, Carles Casamor & Maria Gabás**   Production: **Fundició Dúctil Benito**

Sacharoff is a waste paper bin that combines design, functionality and toughness. The structure is made of cast aluminium with a grainy, matt finish. It consists of a bucket, a lock, and four MB expansion bolts. It can be locked with a key and can be turned around to make it easier to empty.

Arona

Design: **Enrico Marforio** Production: **Ghisamestieri**

Litterbin realized in cast iron UNI EN 1561 consisting of a central body made in one single piece. The design includes the skeleton of the refuse container, the protection cover, the fixed wing and the anchor plates for ground fastening. The cast iron movable wing rotates on the central axle, allowing the wing to be opened to take out the garbage. The steel bottom is welded to the movable wing. The item has a bag-clip ring in galvanized steel. It has a 55-litre capacity.

Litos

Design: **Área de diseño Gitma** Production: **Gitma**

Litos is a strong and solid cylindrical waste paper bin made of artificial stone. It is in fact a multi-purpose item as it can perform equally well as waste paper bin, flower pot or ash tray. It has a concealed system for emptying and is easy to maintain.

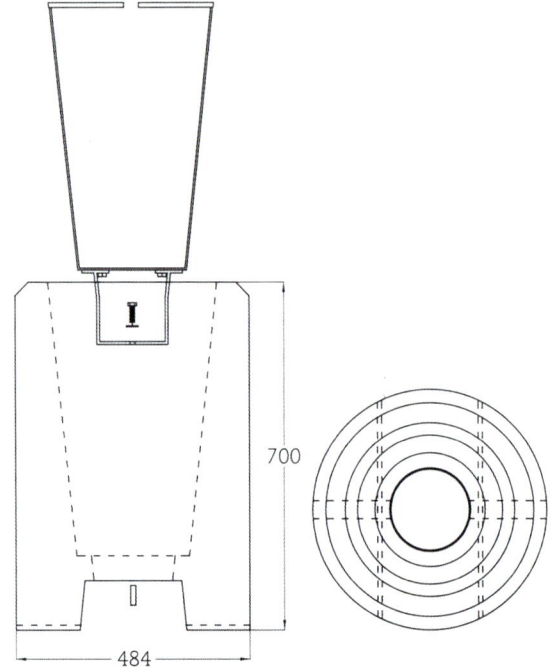

Mobilia

Design: **Erik Brandt Dam** Production: **GH form**

The Mobilia dustbin has been developed emphasizing the expression of simplicity while setting new standards for the work environment regarding how efficiently it can be emptied. The dustbin respects the height and materials of the Mobilia bench, which underlines the wholeness of urban space. By developing the Mobilia series, GH form and Erik Brandt Dam have created an inventory of urban items developed as a set of building components that unify quality industrial products that leave room for individualized layouts according to the circumstances of each project. The Mobilia dustbin is partly buried due to the brief requesting both architectural expression and function; its lower emptying height makes the job easier for personnel. The dustbin's inner space is equivalent to an 80l standard plastic liner, the most commonly used type. The plastic liner is completely concealed in the body, which is easily opened for emptying. The Mobilia dustbin with the item number EBD.10.2022 is made of untreated cast iron. The container is also available with a surface treated design, corrosion class IV, in cast iron or steel.

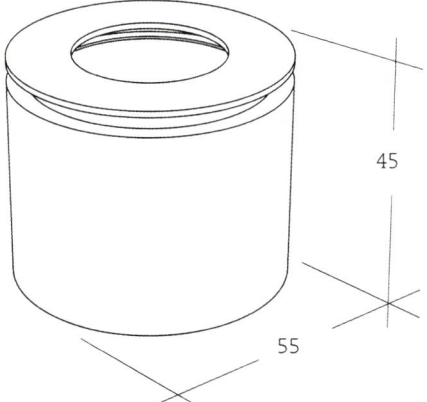

Wastebaskets

Starck

Design: **Philippe Starck**   Production: **JCDecaux Design**

This waste paper bin designed by Philippe Starck combines aesthetic values and functionality. The inverted elliptical body is made of cast iron, completed with a dome-shaped lid with three lateral openings. It is easy to empty thanks to the light-weight inner basket with anatomical handles. It is also easy to clean and remarkably vandal-proof, as it is entirely made of non igniting materials and the inside is only accessible by means of a special key.

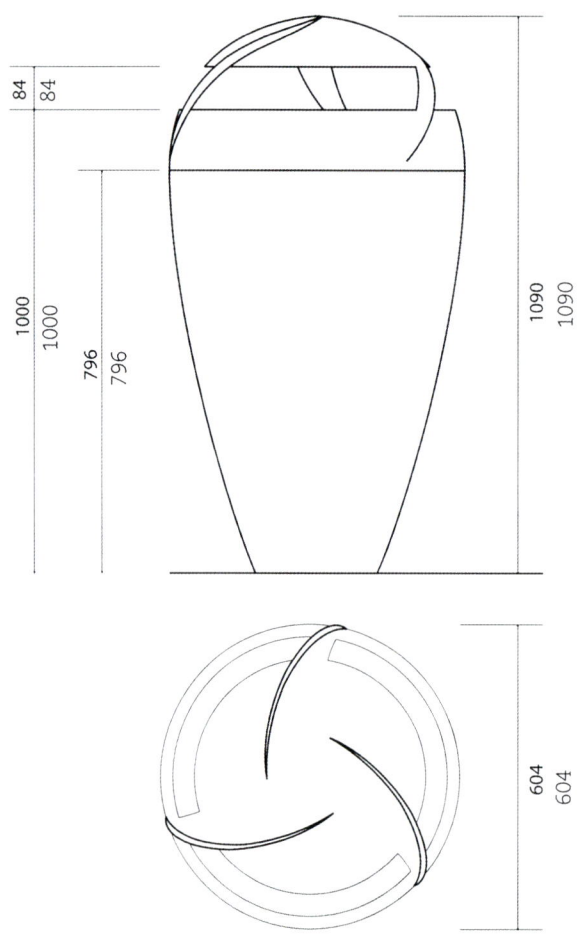

Wastebaskets

Lakeside
Design: **Margaret McCurry**   Production: **Landscape Forms**

The Lakeside wastebasket is part of the Landmark Collection of outdoor furniture created by distinguished architects and designers and inspired by familiar themes in nature, architecture and historic design. Lakeside litter receptacles are available in side or top-opening designs. The side-opening litter receptacle holds 30 gallons, the top-opening version holds 35 gallons. The steel panels are available plain or with plasma-cut decoration of grass or leaves. The standard litterbins come with a removable black polyethylene liner. Standard versions are freestanding or have a surface mounted support. Decorative motifs from nature inspired in trees and grasses are cut into the metal using state-of-the-art plasma-cutting technology. The new Landscape Forms colour palette provides fresh, out of the ordinary hues in powder coat finishes.

Cornet

Design: **Hruša & Pelcák Architects**   Production: **Mmcité**

This distinctive conical litterbin appears to have been stuck into the ground. The original effect is effective on green, paved or even hillside localities. The galvanized steel inner bin is firmly attached to the supporting frame of the bin by a steel cable. The stainless steel body has a galvanized steel supporting structure. The inner bin is made of galvanized steel. The anchoring system remains entirely hidden under the ground.

2278

Design: **Alfredo Farne**   Production: **Neri**

A conical steel column supports this litterbin made of hot-galvanised steel sheet. It has a terminal decorative element in steel. Its standard finish is dark gray. The internal bin is polythene. Dimensions: height 99 cm; width 39 cm; overhang 50.5 cm. Capacity 30 litres.

Paralela
Design: **Juan Cuenca Montilla**   Production: **TD Cabanes**

Paralela is a litterbin the support of which is constructed by means of a "U" shaped mounting bracket made of carbon steel. It has a capacity of 25 litres and can be adapted for use as an ashtray. It is made of galvanized steel plate with rows of perforations, the emptying system is gyratory and the basket is fixed in position with a slam-lock. The entire object is fixed to the ground with bolts and metal expansion-plugs. The finish is hot-dip galvanized, plus a layer of base paint and another of paint.

Rodes
Design: **José A. Martínez Lapeña - Elías Torres**   Production: **Santa & Cole**

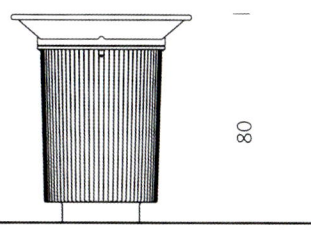
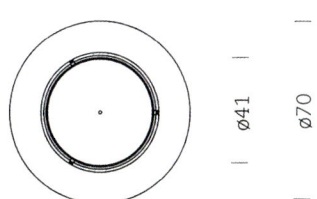

The Rodes litterbin offers great durability and size, and is made of materials that require zero maintenance. The body is made of cast steel with a layer of anti-corrosive treatment and a coat of powder coated black paint. The upper part is of cast aluminium, powder painted in gray. The two pieces are fixed together with stainless steel screws. It has a stainless steel ring inside that holds the rubbish-bag. It is anchored to the ground with three stainless steel bolts that go into previously prepared drill-holes filled with quick-drying cement or epoxy resin or similar materials.

Rambla

Design: **Guillermo Bertólez - Javier Ferrándiz** Production: **Santa & Cole**

This is a robust, elliptical litterbin made of stainless steel, painted steel or galvanized steel. It is very practical and suitable for narrow streets. The urban Rambla litterbin, upright or attached to the wall, is a simple, economic and long-lasting product. It has an elliptical base and a simple, sober appearance and reasonable capacity. It is especially suitable for narrow pavements. Its opacity hides its contents and the unit has remarkable compositional possibilities when placed diagonally in groups, front to back or parallel, to reinforce any particular arrangement. The criteria that determined the design were comfortable use, easy handling and aesthetic appearance.

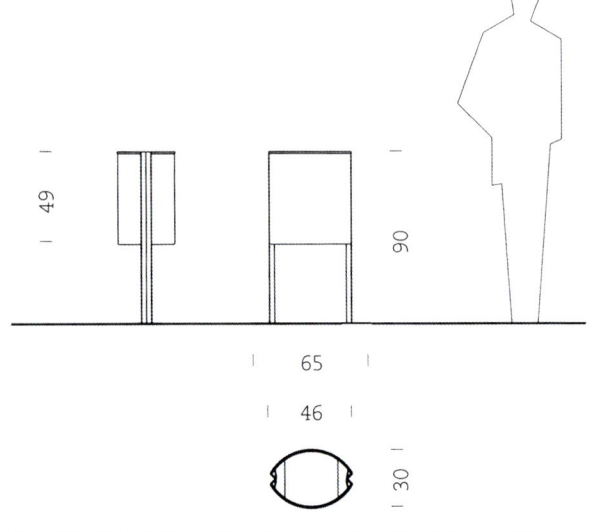

Cordillo

Design: **Greg Healey** Production: **Street and Park Furniture**

The impetus for this furniture was a trip through South Australia's outback. Greg has responded to the forms, textures and colours of this extraordinary landscape in the realisation of much of this new range of urban street furniture. This adaptable range of furniture is particularly suitable for use in contemporary streetscape design, as it draws on Australian landscape forms in a unique and innovative way.

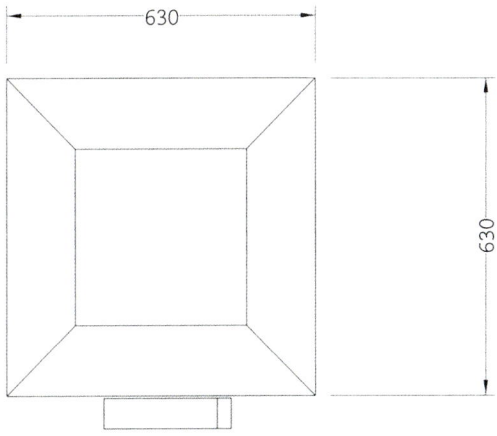

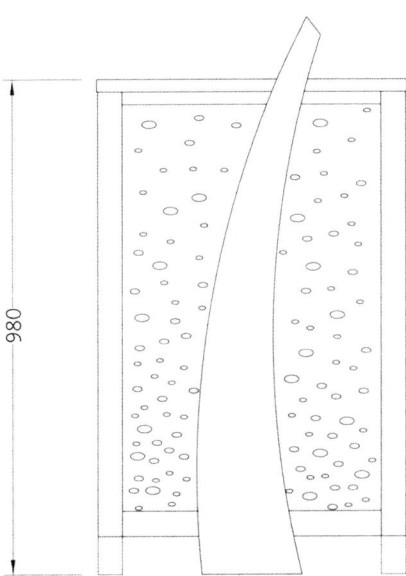

Wastebaskets

Radium

Design: **David Karásek, Radek Hegmon**   Production: **mmcité**

The Radium litterbin is a geometrically styled body with softly rounded edges that is the direct result of the technology of bent steel sheet. The support structure is a rectangular ring with rounded edges that carries the inserted box, which in turn contains the galvanized steel inner bin. The complete piece is an elegant and durable example of public furniture. Its design establishes an associative link with the Radium benches and bollards. The supporting frame is made of galvanized steel, with a rustproof covering and painted in a standard shade. There is a forward opening lockable door and an ashtray with a stainless steel cigarette extinguisher integrated into the piece's top cover. The item is designed to be fixed to a concrete base or a pavement by means of stainless steel anchoring bolts.

Valet

Design: **David Karásek, Radek Hegmon**   Production: **mmcité**

This slim and inconspicuous freestanding ashtray related to the SL bollards range is suitable for various localities. The article consists of a steel angle iron post that bears a lockable stainless steel box for cigarette ash or plastic bags for dog excrements. Its function is made identifiable by a cigarette symbol, engraved on the steel and enamelled red to increase its visibility. Another model serves as a support of the plastic bags for dog excrements. The galvanized and powder coated steel body rises 1000 mm (39,3 inches) over the pavement. Two widths are available, to gain adaptability to the proportions of a given installation: L 60 mm (2,358 inches) and L 80 mm (3,144 inches). Its robust structure and easy manipulation are the major advantages of this modern element of street furniture. It has a baseplate designed to be fixed in the concrete under the pavement, guaranteeing the object's stability.

Caos

Design: **Tobia Repossi**   Production: **Modo**

Litterbin is composed of a 22 mm gauge steel spring that spirals around the body of the bin, which it supports. The bin is made of metal plate. The item stands upon a sandblasted concrete base. It is easy to empty as the bucket is extractable.

Cinderello

Design and production: **Modo**

Ashtray and support post for outdoor use. The crossing of a curved plate with a 10 mm ø tube creates the main structure, which is fixed to the ground. The ashtray is a stainless steel moulded plate, with curved edges. The inner grating to put out the cigarettes is made of AISI 304 stainless steel punched plate, designed to be easily laid and inserted into the main structure.

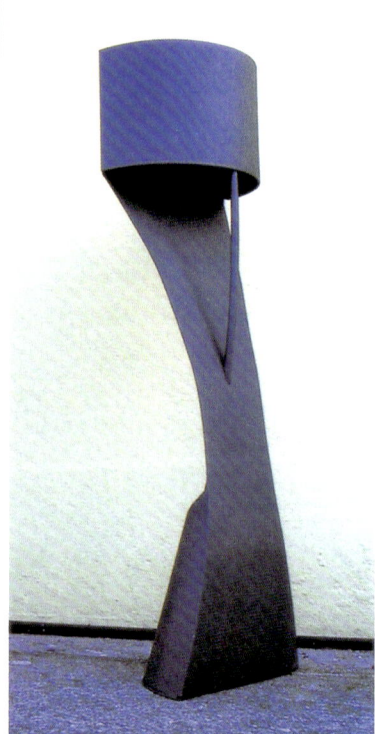

Pandora

Design: **Alessandro Riberti – Studio Italo Rota & Partners**   Production: **Modo**

A litterbin with a cast iron body (also available in aluminium on request), and a cast aluminium lid with three inlets, in this version with simple holes; it is provided with an inner stainless steel ring that blocks the liner, unlocked with a special safety key. Designed for use with an inner galvanised steel basket or dustbin liner. An inner stainless steel ashtray is available on request. The surface is treated with anti-rust protection and covering with thermosetting powders in silver, charcoal grey and metallic bronze.

Vega

Design: **Alessandro Riberti**   Production: **Modo**

A litterbin consisting of a 3 mm steel plate skeleton with a bin liner ring and a 1.5 mm turned steel plate frontal case lined with 70x28 mm impregnated iroko or pine wood slats. The openable hatch has revolving bodywork to facilitate removing the liner. The key closure is triangular.

Fountains

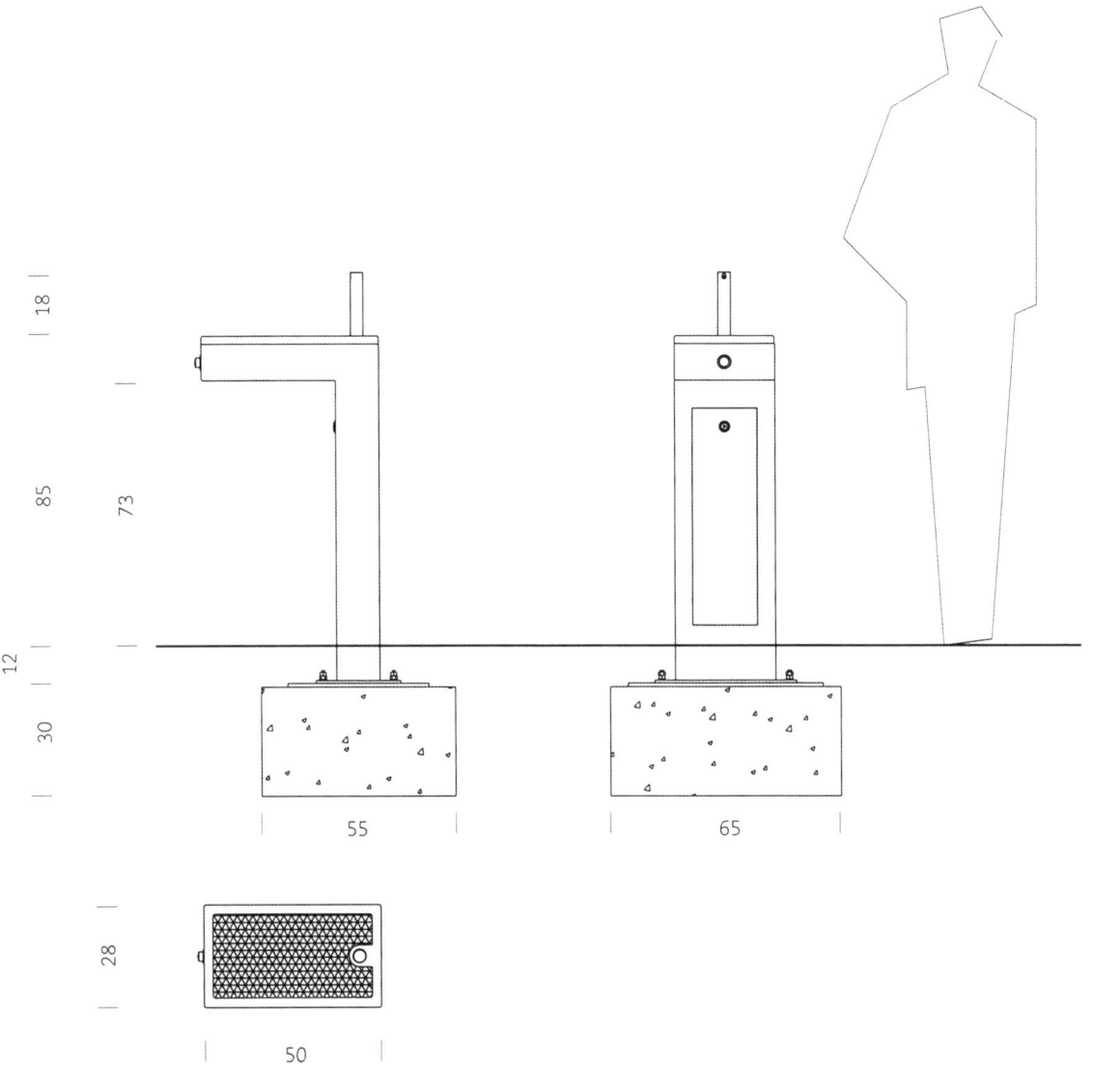

Periscopio

Design: **Diana Cabeza**   Production: **Estudio Cabeza**   Development team: **D. Cabeza, A. Venturotti, D. Jarczak**

The Periscopio fountain is a water-dispenser with a grid and a drainage system, designed for public space in the city. It is designed to complement the Rehué benches, likewise created by Diana Cabeza. Periscopio is made of cast iron with a powder-coated finish of transparent thermo-convertible polyester. It is easy to install as it is simply inserted into the concrete floor.

Lilla

Design: **Tobia Repossi**   Production: **Modo**

The structure consists of 60 mm diam. stainless steel tube, sandblasted and shaped according to a parabolic curve. The fountain has a stainless steel supporting base plate, with laser carvings for water drainage; at the base it has a 250 mm diam. stainless steel bowl. Water outlet is available in the upper and in the lower part through a timer button at the final part of the tube. The bowl at the base can be used as a drinking trough for animals. Fixing to the ground requires sinking the drainage basin in a concrete foundation at a depth of 220 mm underground, which corresponds to the height of the basin.

Caudal

Design: **Pau Roviras y Carlos Torrente** Production: **Santa & Cole**

A public fountain based on geometric forms. Original, simple and pleasant, especially designed to be comfortably used by any citizen. Developed and improved in cooperation with Barcelona City Council and organisations of disabled people, it is accessible and suitable for everyone. Easy to operate, it was designed thinking of minimum consumption with an appropriate and constant flow to facilitate its use by people with motor disabilities.

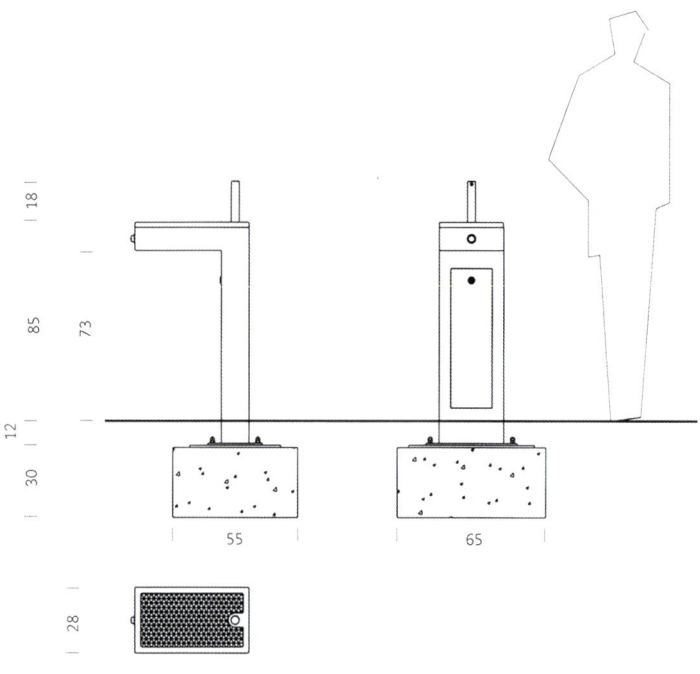

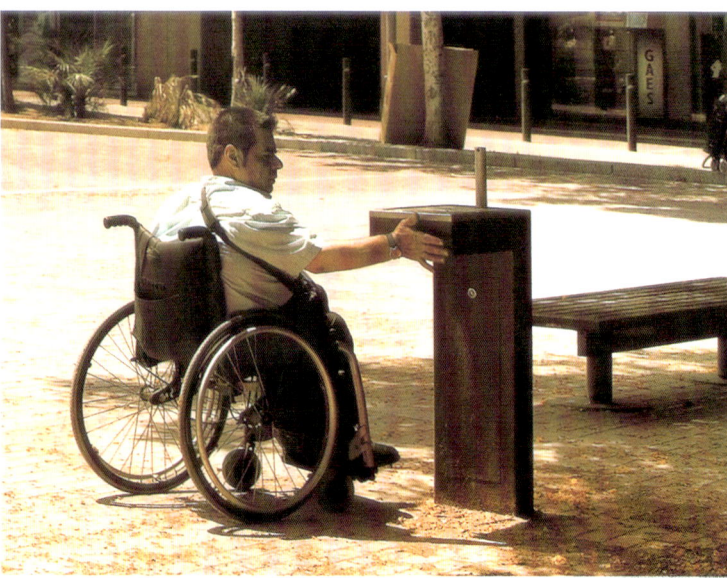

130   Fountains

Estena

Design and production: **Tecnología y Diseño Cabanes**

The Estena drinking fountain is an example of perfect adaptation either to a fresh and modern context or to a site with important historical connotations. Its recycling capacity is of 30 litres. The base is made of stainless steel faced with Corten steel. The outer layer is of transparent paint.

Fountains

Lights

Faf

Design: **Joan Forgas**   Production: **Alis**

The Faf streetlight series is a set of urban lighting elements. The series includes a light for the road, available in two different heights and at different angles and a path light, available in three heights and at different angles. The alternation of the two produces an undulating visual effect. The lights are formed by a 180 cm steel lower post, which incorporates a covered slot for electrical connections and security lock. This allows access to the connections box. Likewise the front section of the post has a plate soldered onto it that serves to attach the light to the pavement. The post is hot galvanized steel and powder paint finished in epoxy RAL 8008, to which a layer of anti-rust and anti-grafitti paint is added. The upper section of the post is a steel tube. The two posts are joined by way of six nuts and bolts. The light fixture is attached via a 5 mm thick fork-shaped steel piece, which is screwed to the upper post.

Lausanne

Design: **Carles Valverde**   Production: **Alis**

The work of Carlos Valverde is nourished by the most contemporary trends. One of its characteristics is that it changes according to the space at hand and by the uniqueness of each one of his designs. In this case it is a sober piece of urban sculpture, monumental and elegant, created to resolve a problem of space in a particular location by the recurrent use of Corten steel.

Mota

Design: **Iñaki Alday, Margarita Jover, Maurici Ginés** Production: **Escofet**

The design by architects Iñaki Alday/Margarita Jover and the lighting designer Maurici Ginés from ARTEC3 was realised as an initial series of 880 concrete boundary marker/post light units with a built-in light, manufactured by Indalux, to illuminate the walkways of the River Park at Expo Zaragoza 2008. The name Mota was taken from the sixth meaning given in the Dictionary of the Royal Spanish Academy: "A low elevation, natural or artificial, standing alone on a plain." The place where it has been installed certainly fits the definition.

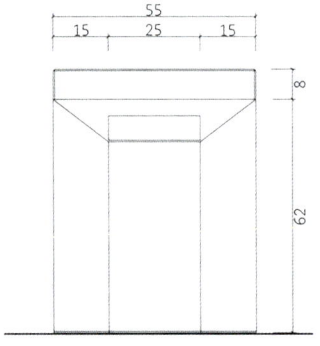
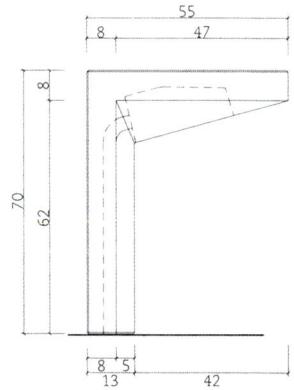

136   Lights

Romana

Design: **Joaquim Carandell**   Production: **Fundició Dúctil Benito**

Romana is a bollard designed with a double function: security and decorative illumination. It is a useful signalling element for pathways, parks and pedestrian areas, with a minimalist design that features essentially purist outlines enhanced by a zincified steel finish.

Sidney
Design: **Joaquim Carandell**   Production: **Fundició Dúctil Benito**

This urban illumination system features a design that stands out for its solidity and sturdiness. It is adequate for the illumination of highways, industrial estates and public areas. The body is made of hot-dipped galvanized steel with a wrought-iron finish combined with gray.

Design: **Francisco Providência**   Production: **Larus**

The angle of inclination gathers the components of the lighting device: the vertical elevation of the pole and the arm with horizontal projection. The illumination is housed inside the tube and projects light vertically. The tube features a cut-out where the cylinder is joined to the illumination cone. The design is inspired in a reductive and laconic expression, suggestive of freedom. The body of the element is made of stainless steel and it carries a 150W metal halide lamp, colour 3000 k, 1531 lux and 13 500 lumens.

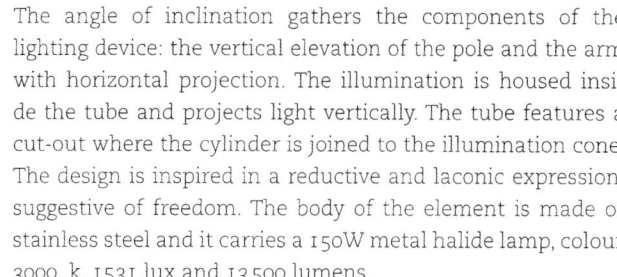

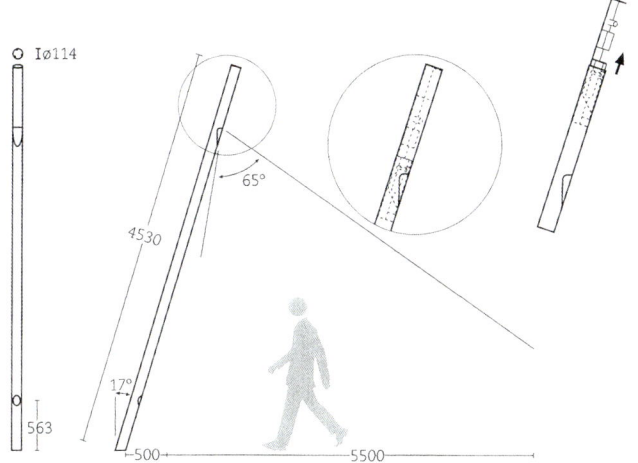

Lights

Ipsilon

Design: **Seste**   Production: **Ghisamestieri**

The Ipsilon lamppost is characterized by the originality of its outline and the materials implemented, wisely used and put together in a harmonious union of shapes and proportions. The system consists of a U-shaped component in drawn aluminium and another in circular section, both having a 170° bend in the upper portion, to house the luminaire. The Ipsilon lamppost, available in two models, consists of a drawn aluminium element in U-section, with maximum dimensions 130 × 150 mm. The channel is 6800 mm high and has a 70° bend. In the double version a second element is joined, circular in section, 6800 mm high, in drawn aluminium, with the same curvature as the first. The connecting elements are realized in aluminium tubes Ø20 mm inside which there is a stainless steel tie M10. The luminaire is in die cast aluminium with a tempered flat glass screen. Ipsilon mounts asymmetric reflectors and metal halide, sodium vapour and mastercolour lamps from 70 to 250 watts. The Ipsilon system can be utilized for lighting medium and large sized avenues and squares, historical centres and residential areas. The light flux Ipsilon emits to its upper hemisphere is nil, in conformity with the strict regulations concerning light pollution.

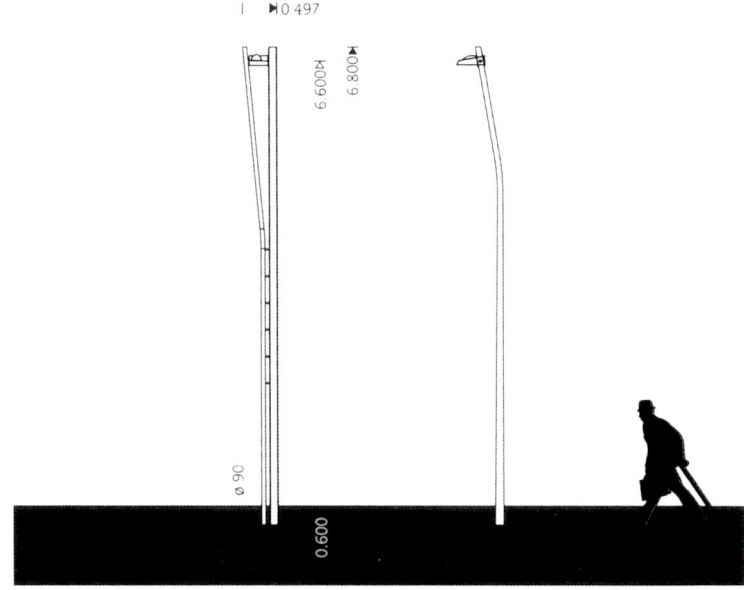

Jupiter

Design: **Emilio Ambasz** Production: **Ghisamestieri**

Jupiter is proposed as a roadway lighting element of great visual impact. The pole consists of a base decorated with red circular ornaments and three oval section aluminium extrusions joined together by means of a connecting system. A drawn aluminium tube characterizes the upper part with slots lit from the inside. This pole is extremely light and so elegant that it enhances the quality of any specific background, urban or suburban. The base, in hot-galvanized steel, is circular in cross-section; fifteen (15) thermo-rubber collars are inserted into the pole base; three elliptical elements in drawn aluminium set at 120° are joined to the central pole by means of stiffening beads; the central element in drawn aluminium is circular in cross section, where three indentations have been made to house decorative strips of coloured LED lights; the upper ferrule is in cast aluminium; the steel bracket is Ø60 mm in diameter; lighting body is in cast aluminium with an IP66 optical compartment. The Jupiter is suitable for lighting medium and large sized avenues and squares, historical centres and residential areas. The light flux emitted by the Ipsilon system to its upper hemisphere is nil, in conformity with the strictest regulations concerning light pollution.

Lights

Volcano

Design: **PLH Design**   Production: **Louis Poulsen**

Volcano is designed as an on-ground fixture. The product is mainly suited for image and representative areas due to the fragility of the material. The idea behind the product is a cone, which reflects the light from a reflector placed in the top. When lit the shape of the cone appears as a light profile and makes the light seem weightless. It gives the impression that it has emerged from the landscape, hence its name. The light is directed from the top reflector on to the surface around the Volcano. This causes the Volcano to emit symmetrical light and makes the fixture suitable for indication and landscape lighting.

Skot

Design: **Lauritz Knudsen**   Production: **Louis Poulsen**

Skot is a traditional maritime design, a simplification of the bulkhead lamps on ships. The name is a Danish word for bulkhead. The lamps are used both outside and inside ships. They were protected by a cross or a metal net, which in fact hardly affected the light distribution. This type of light fixture is still used on ships, but due to the influence the maritime context has on fashion, these fixtures have been used in a lot of situations requiring a robust lighting fixture or wanting to suggest a maritime environment. The original design is more than 100 years old. Skot Bollard provides both direct and diffuse lighting with a choice of either clear or white opalescent diffusers. The half-masked crossguard accentuates the downlight.

Nawa

Design: **Antoni Arola**   Production: **Metalarte**

The Nawa lamp comes in two forms, one is a wall-hanging model and the other is a free-standing element. The structure is made of extruded aluminium, finished in silver lacquer. The light diffuser is made of opalescent methacrylate. The body is 8 cm in diameter and can rise up to 250 cm. It requires an anchoring system.

Panamá

Design: **Mario Ruiz**   Production: **Metalarte**

This illumination system consists of two simple components, a cylinder and a rectangular prysm with rounded corners. The designer has carried out a dug-out exercise, to eliminate all the unnecessary mass until he had achieved the lightest possible elements to house the sources of light. This series of balanced luminous objects features gentle, rounded profiles, creating an item designed for outdoor areas that intends to avoid the classical tubular beacons. The light source remains hidden, to fulfil a double function: light is projected directly onto the ground while the illuminated edge of the lamp provides the visual signal required.

Lights

Diorama

Design: **Ramón Benedito** Production: **Santa & Cole**

Ramon Benedito has created an attractively-shaped lighting unit, inspired on traditional lines but made with modern-day materials and technology and destined to last because of its pleasant shape. A highly familiar street lamp that looks as if it had always been around while at the same time telling us something new. It is made of cast aluminium, has a plastic diffuser and is equipped with compact fluorescent lights. It is mounted on a cone-shaped matt aluminium column. The light projection angle reduces its light pollution ratio to zero.

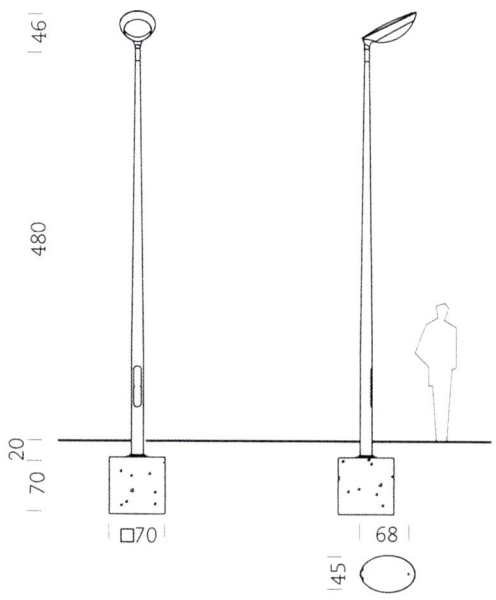

146 Lights

108

Design: **Enric Batlle - Joan Roig**   Production: **Santa & Cole**

A simple and functional street lamp, which arises at an angle. Its shaft has a continuous rectangular cross-section, designed to be useful without prominence. The tubular element has a rectangular cross-section, finished in hot-galvanized steel. The hot-galvanized steel tube column and arm have a continuous cross-section of 120 x 280 mm for the 4.7m lamp, and 150 x 300 mm for the 7.5 m lamp. The arm leans 15 degrees in relation to the horizontal. Optionally, they can be delivered painted. Optical unit and extrusion reflector in high gloss anodized aluminium and tempered glass diffuser, equipped for discharge lamps, metal halides or high pressure sodium vapour (max. 150 W). Optionally, double level equipments can be included for the command line.

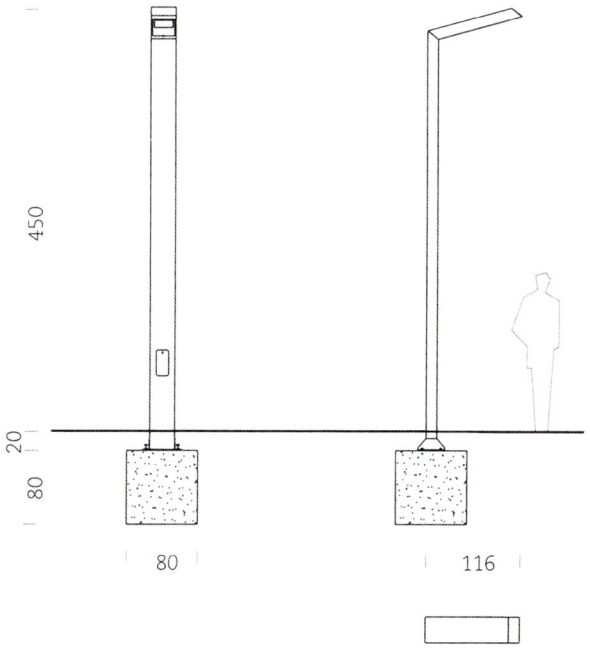

Lentis
Design: **Alfredo Arribas**   Production: **Santa & Cole**

Alfredo Arribas has designed a modern street lamp with a classic lenticular shape. Thus, he enlarges the projection of light of the classical globe in the bottom half and avoids light pollution with the opaque top half. Alfredo Arribas seeks a classical (lenticular) shape to create a lamp with a modern flavour: the result is the Lentis street lamp. On flattening the dimensions in relation to the classical globe, he increases the area of diffusion and its features. A reflector on the upper half prevents light pollution and increases the performance. The support is a classical cylindrical shaft which leaves full prominence to the lamp. It is a streetlamp adjustable to any kind of urban space. To create the Lentis streetlamp, Arribas found inspiration in the Copa stool, by the same author and very much used within the last twenty years in bars and restaurants. In the late 8o's, Arribas was busy designing bars and nightclubs that impressed everyone and established Barcelona's modernity.

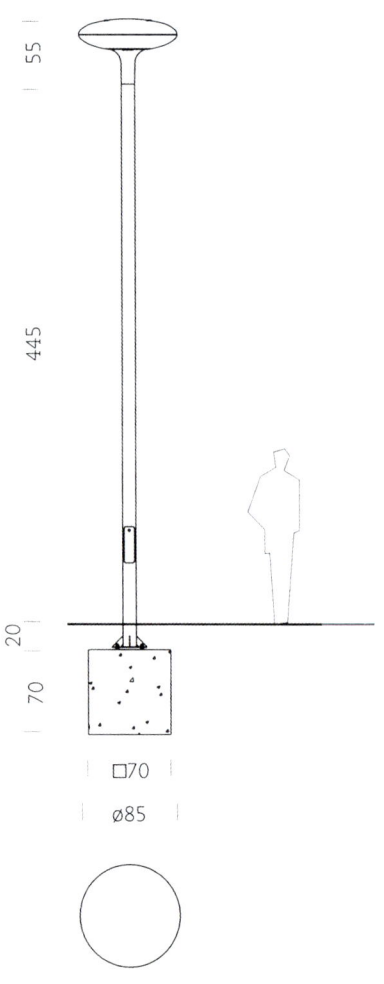

Palmería

Design: **Antonio González Cordón** Production: **Santa & Cole**

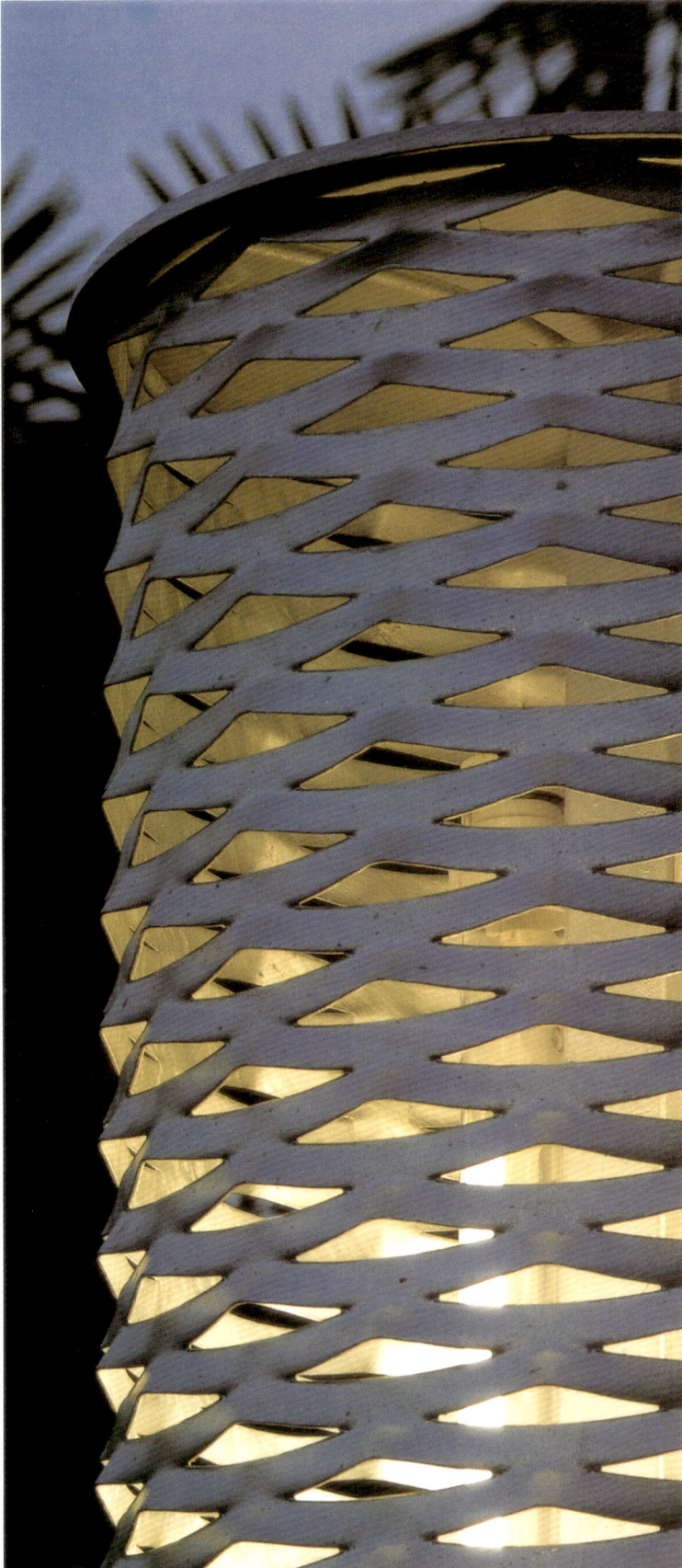

A palm tree trunk, an easy form to recognize, inspired the author in creating this street lamp that emerges from a palm grove like a light totem, symbolizing the union of nature and technology. Its rational and sober silhouette, a one-piece cylinder of hot galvanized "deployé" steel sheet, encloses two fluorescent lamps that create a relaxing and mysterious atmosphere. The idea originated in Almeria, in a large palm grove made into a public garden. Where palm trees formerly stood, these totems, now erect and shining in the dark, blend in and enlarge the area. The result is a forceful, elegant piece. This street lamp is designed as a tall marker that brings personality to the environment, shedding out a soft warm light to accompany us in the night, without imposing its presence.

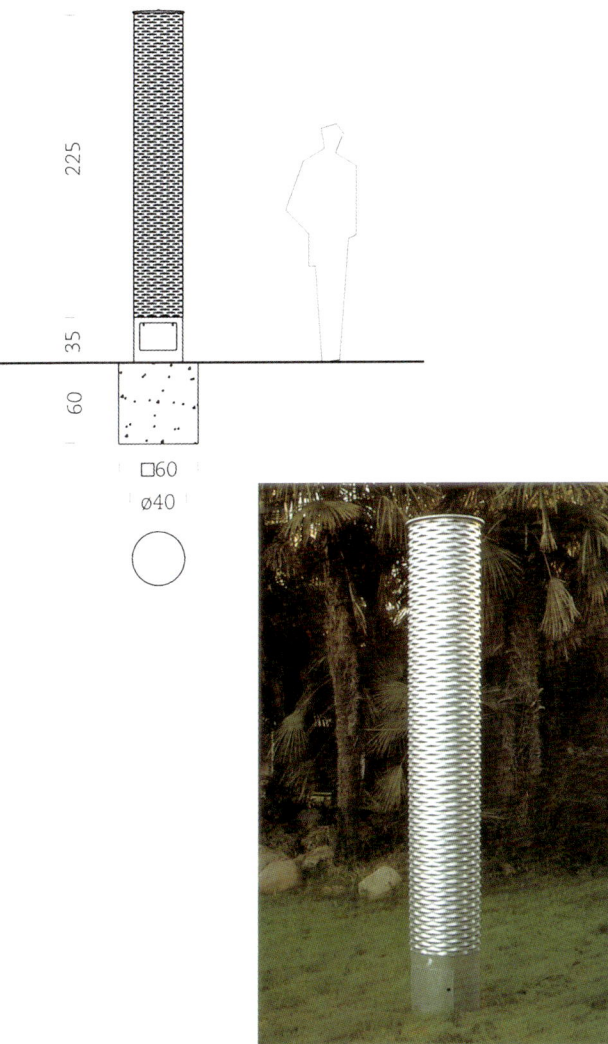

DL 10
Development: **Siteco**

With this pioneering performance the DL 10 demonstrates how outdoor lighting in the future will look, an organically flowing form that ascends from the mast spigot to the luminaire head. The aim of the DL 10 was to implement LED technology to satisfy the demands for the illumination of prestigious squares, roads and city centres in accordance with current standards. The result is impressive. The first Siteco road luminaire incorporating LED technology represents a new luminaire generation, combining state-of-the-art LED technology within an innovative and functional design. The DL 10 provides lighting according to standards for roads and squares with a pleasant white light or coloured accent lighting for the first time with one luminaire. The construction of the luminaire body gives the impression of coming from one casting and is attractively graceful. This light impression is based on the creative potential of the technology used: LED lighting technology allows completely new housing forms and low construction heights – DL 10 has taken good advantage of this bonus. The materials, diecast aluminium with Siteco metallic grey coating as well as an optical enclosure of brilliant, part-matt PMMA that flushly fits into the graceful form rounds off its harmonious appearance.

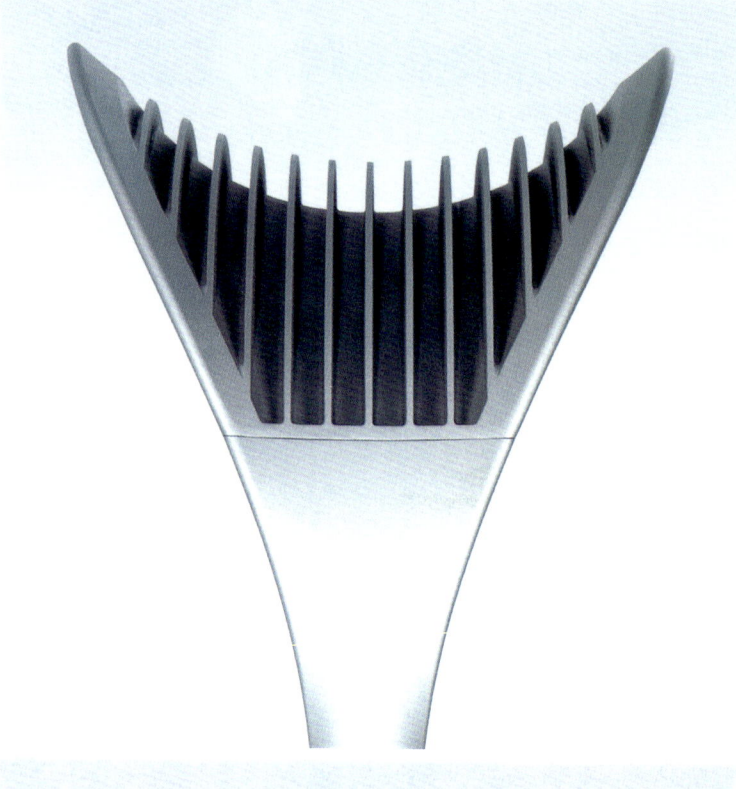

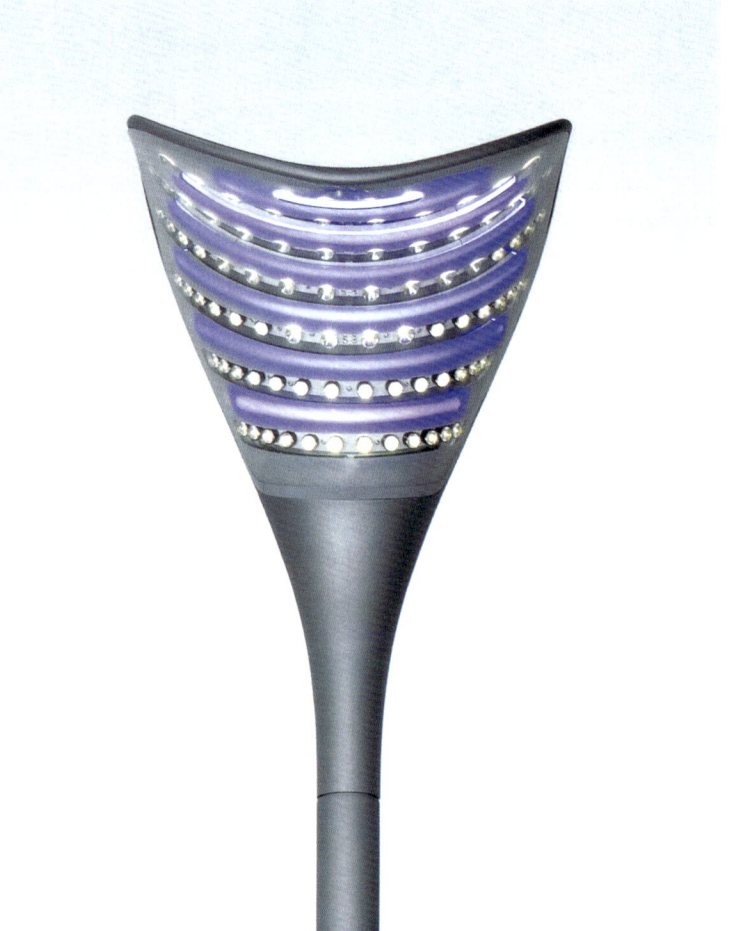

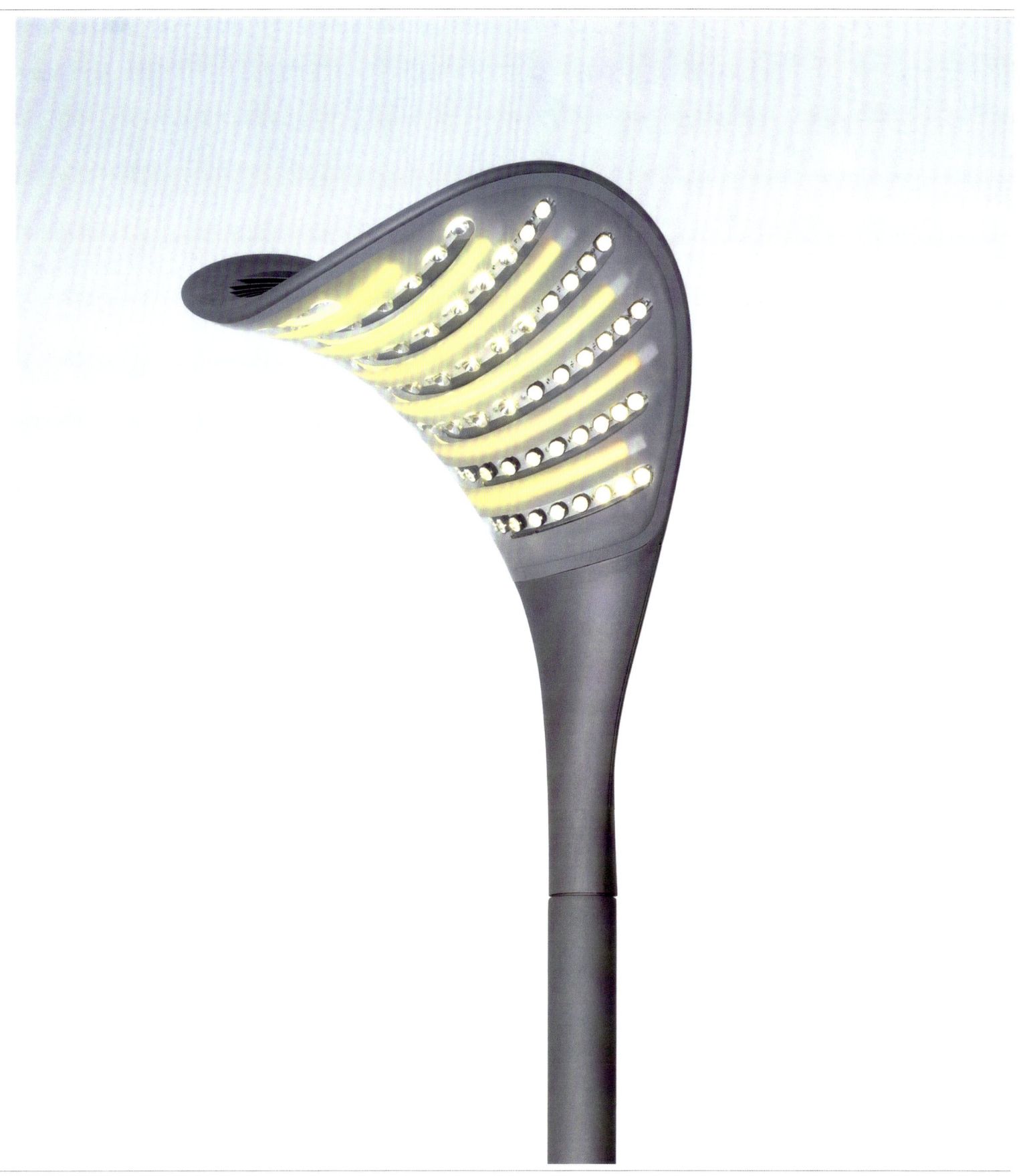

Solitaire

Design: **Licht Kunst Licht**   Production: **Siteco**

As the Washingtonplatz in front of the Berlin main railway station acquires its final appearance, a single light pillar will illuminate the area. The Berlin railway station has been running now for a long time, but around the glass palace construction continues: with Europaplatz to the North and Washingtonplatz to the South, two important railway squares are being created. Constructed according to designs by landscape architects Martha Schwartz and Gabriele Kiefer, the concept consists of an open site defined by steps and bordered on the West by a grove of trees. The Café Pavilion on the lower part of the square will be replaced by a 40 metre high cube designed by the now deceased Oswald Matthias Ungers. With the slimline pillar luminaire erected near the entrance to the station, the Washingtonplatz already features one creative highlight. The 26 metre high mast constructed of centrifugal concrete is crowned by a 5 metre high light module whose six slots can accommodate up to 60 projectors. The solitary unit is to be supplemented with the addition of two more pillars.

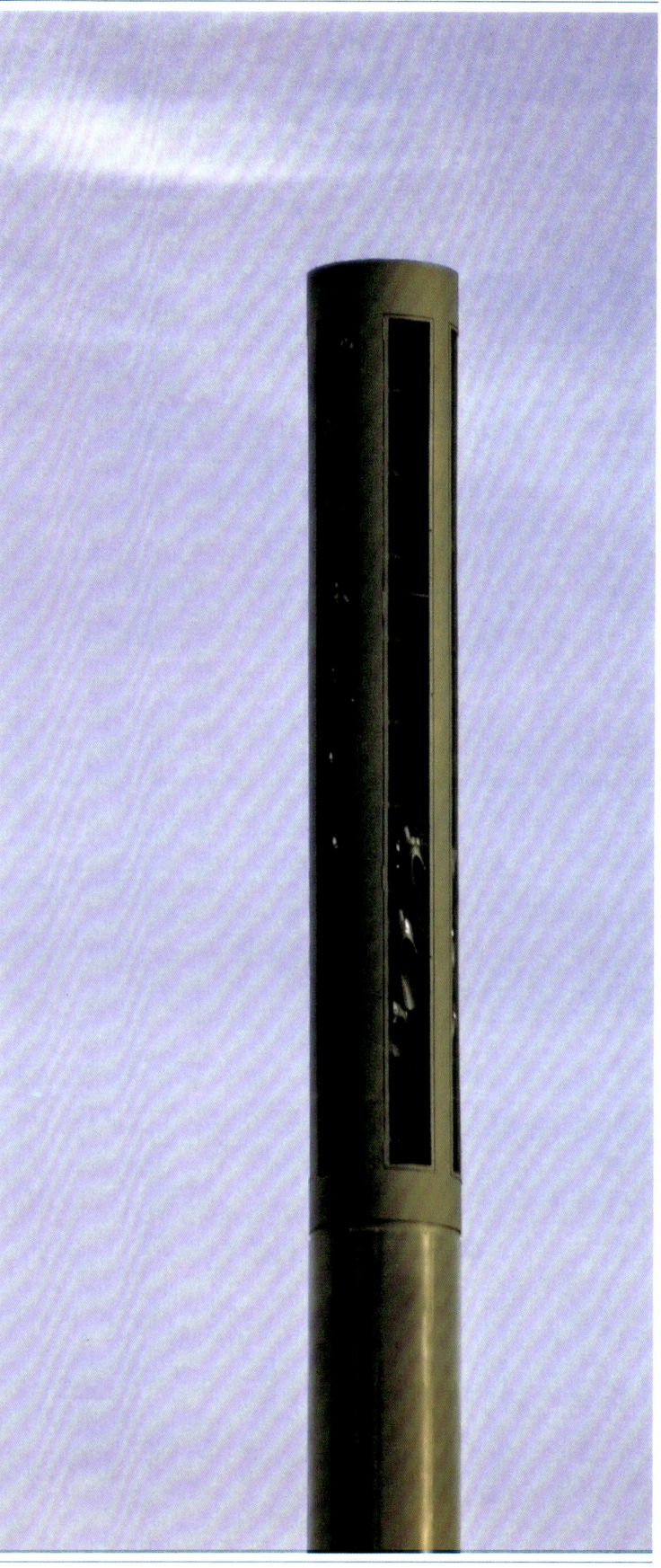

Lights 153

Gropius
Design: **Eduardo Albors**   Production: **TD Cabanes**

Gropius consists of a conical post, a head or screen and a terminal (optional), also conical and finished with an item made of injected plastic that houses a small pilot light. The platform of the lamp is fixed to the post and houses a glass diffusing screen and a watertight join. The piece that houses the pilot-light can be red, green, blue or yellow. This little light is a differential accessory added to create a linear play of light when a number of the lamps are installed in a row. These fixtures can be finished in silver gray, graphite, a combination of both, or matt white. The post can be of stainless steel. Also available is an unfinished model without the pilot light, which is then substituted with a watertight lid.

Java 2

Design: **Javier Valverde** Production: **TD Cabanes**

Java 2 is an illumination system composed of three bodies, square in cross-section, installed at a 90° angle to the horizontal surface of the ground. The luminaire is located inside each body of the lamp, where it is protected by a shutter at the top. The columns or posts are designed to illuminate a wide perimeter in an expansive way, creating a pool of light in the shape of a clover-leaf. The lamp can be provided in stainless steel, galvanized steel or even Corten steel.

Tres

Design: **Marta Ferraz & Paula Cabrera** Production: **TD Cabanes**

The triangular geometry of Tres is capable of illuminating the whole perimeter expansively. The illumination is produced by an ice-coloured strip of polycarbonate 20 mm thick, in each side of the object. This, added to its formal subtlety, gives the lamp a sober profile, but produces a special rhythm that is ideal for defining an area or marking out a direction or path. The shaft is made of galvanized steel painted in a range of colours, available on request. It houses a low energy electronic bulb for the beacon and the lamp uses three 1500mm long 35W fluorescent tubes. The lamp is anchored to the ground by bolts and special security plugs embedded in the pavement.

Light Stripe

Design: **SLA**   Production: **GH form**

GH form and SLA have developed a beacon light system in cast iron, the Light Stripe. With the Light Stripe, it is now possible to create modules of up to 30 illuminated meters in length. The Light Stripe is designed to be built into the pavement as a total concept. The starting point of the Light Stripe is located in the kiss and ride space in the city centre of Frederiksberg, where GH form have delivered the cast iron elements. The system consists of a cover plate in cast iron, with LED light spots cast in clear acrylic. For routing of LED light cables and mounting of cast iron elements, a groove in polymeric concrete is used.

Hom

Design: **Alessandro Caviasca** Production: **SIARQ**

Hom is a photovoltaic lamp of tremendous output. The panels transform solar intake radiation and store electrical power, obtaining a highly efficient illumination system. Each Hom lamp uses 30 LED bulbs of the most recent design. The lamp generates no heat, requires no maintenance and provides illumination throughout the entire night, even in the wintertime. No trenches have to be dug to install them; they can be placed in any out-of-the-way location, far from the electrical mains. The lamp has a useful life of 50000 hours. The design of the lamp represents a human figure that embraces the light of the sun and defines its profile. Hom is ideal for spaces of large dimensions and it saves 255.5 kW/h per year. This translates into a reduction of 153.3 kg of emitted CO_2 with no loss of light power.

Lights 157

Spiral Light

Design and production: **Street and Park Furniture**

Street and Park Furniture were provided with details on the type of light fitting required for the project and the designers had to devise a way of fixing the lights to the pole. To create an aesthetically pleasing detail that would give some flair to an otherwise uninteresting pole, they successfully managed to create a spiral out of a flat bar which was attached to the pole and also allowed access for the wiring. Street and Park Furniture are committed to the view that better streets and civic spaces are a vital part of community well being and endeavour to be perceived as a supplier of street furniture for these applications. They aim to provide specifiers and customers with the convenience of a single supply source for most urban projects incorporating street furniture and the support of specialised experience.

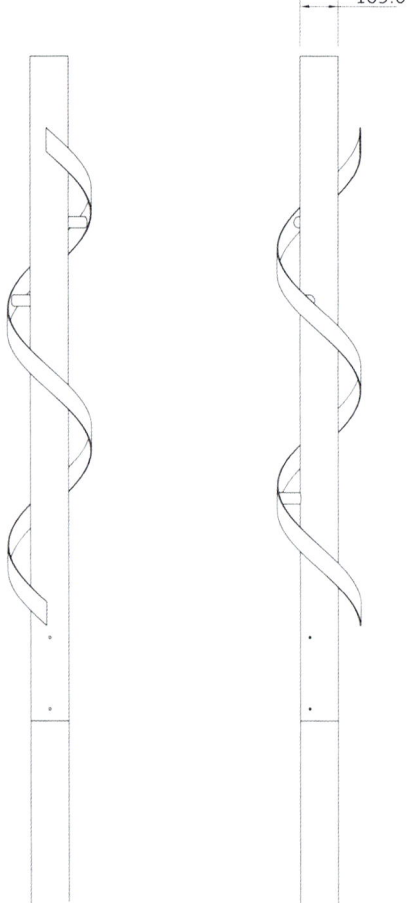

Icon Mini Opal
Design: **Mads Odgård** Production: **Louis Poulsen**

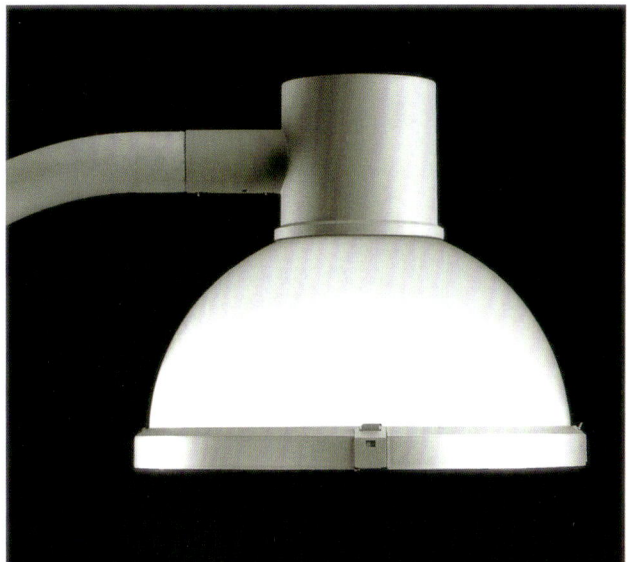

Icon Mini Opal effectively provides cut-off street lighting with less than 2.5% light distribution above horizontal level. The clear, simplistic design is intentional, since Icon is designed for use in numbers. It combines two geometric forms: a cylinder and a hemisphere. Because of the faintly lit hemisphere, the shape is uniform both night and day. Numerous light source positions in the reflector can make both symmetrical and asymmetrical light distribution possible. Different mounting solutions are available to facilitate optimum positioning and meet individual requirements.

Heinola Reading Lamps
Design: **Vesa Honkonen**

Heinola is a town some 130 km north of Helsinki, Finland. Late summer 2004 the Heinola authorities contacted the designer to see if they could use the lighting solution he had created for Raisio, another small town in Finland. Raisio light net, completed 2002, got good response from international press, which raised the image and status of the town. However, Vesa Honkonen had promised Raisio he wouldn't use the same solution in Finland for 7 years, so Heinola was told they would have to be content with something created just for them. They wanted it for a small library plaza; a relatively small urban space, which the designer felt, required some movement. Vesa Honkonen started to dream about the reading lamps stepping out of the library to dance in the street, growing bigger, taking over the street. Having landed in new territory, they started to look over their shoulders, look around, bend down. After a while they calmed down. They started to freeze as they slowly accepted their form and positions. The time for movement was over. They were satisfied and at peace. Their frozen movements took shape as three individual forms. The lower part of the pole is similar in each variation. Three different curved parts permit the fixture to bend in different directions. The lamp head is similar for each variation so three unique fixtures resulted and a total of fifteen units were installed. It was clear form the start that the light source would be metal halide in order to project strong dots of light onto the street and avoid even lighting. It was planned to be a good lighting designer's nightmare, uneven lighting with strong contrasts. Each lamp casts sharp beams onto the street, revealing the shape of the light roaming out of its body. The manufacturer's first model provided extra information about the shape of light, so it was decided to push the lamp deep inside in the body of the fixture to avoid glare. Thus, light bursts out of a black hole in different directions.

Lights 161

Triangel

Design and Production: **Philips**

Triangel combines a strong, easily remembered design with a robust and durable construction. Its characteristic lit triangle at the top creates an optical guidance and secondary lighting. It is designed for use in residential areas, shopping areas, car parks, footpaths/pedestrian areas and other urban spaces where its high efficiency reflectors will provide economic lighting. Tempered safety glass makes Triangel vandal-resistant. The housing is made of robust, die-cast aluminium, painted in black or grey (RAL 7021). The cover is made of tempered safety glass - frosted (GF) or clear (GC) - mounted in an aluminium frame, painted in the colour of the housing. The article weighs approximately 12 kg (26,5 lb). It is suitable for a wide variety of lamp types. Mounting brackets are available to permit single, double or quadruple post top mounting.

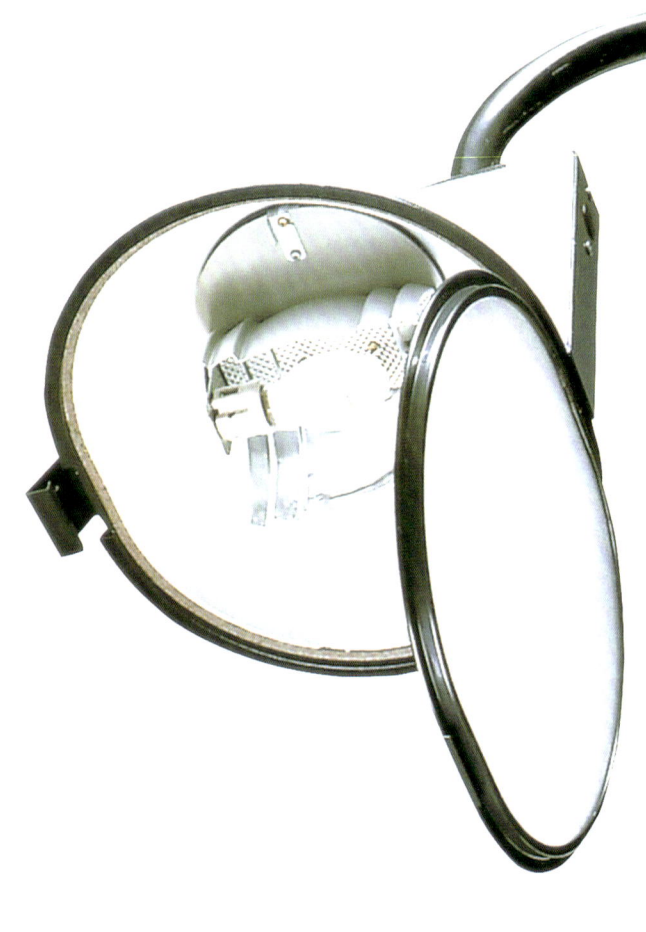

Tournesol

Design: **Philippe Starck**   Production: **JCDecaux**

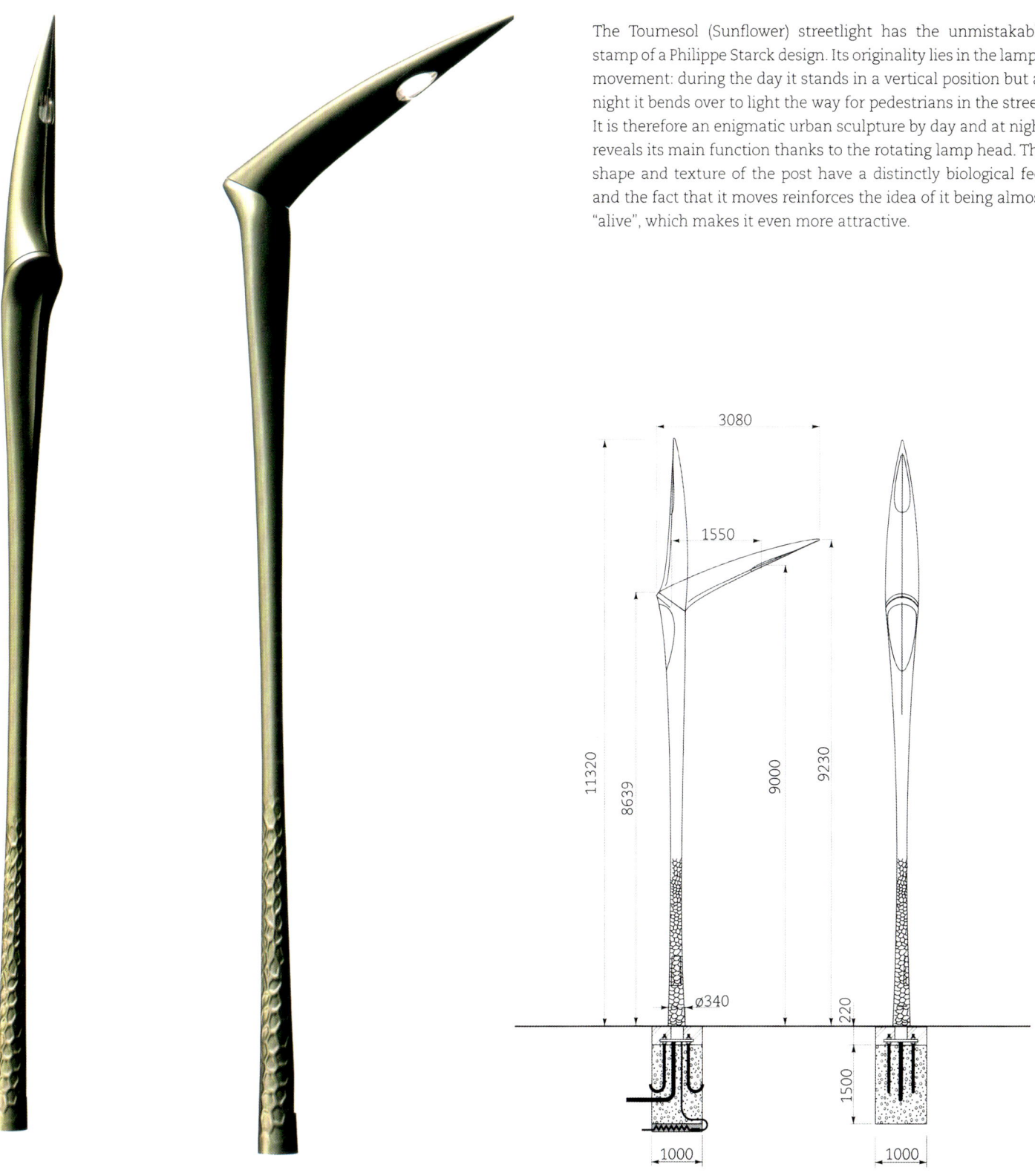

The Tournesol (Sunflower) streetlight has the unmistakable stamp of a Philippe Starck design. Its originality lies in the lamp's movement: during the day it stands in a vertical position but at night it bends over to light the way for pedestrians in the street. It is therefore an enigmatic urban sculpture by day and at night reveals its main function thanks to the rotating lamp head. The shape and texture of the post have a distinctly biological feel and the fact that it moves reinforces the idea of it being almost "alive", which makes it even more attractive.

Rondò

Design: **Piero Ravaioli**   Production: **Ghisamestieri**

The suppression of road crossings, gradually transformed into stream roundabouts, has induced Ghisamestieri to design the Rondò system as the answer to the need to find a pleasant way to illuminate and decorate these new urban areas that are now integral parts of any urban landscape. Its harmonic light shape and the chromatic central obelisk allow for the creation of different settings and solutions, in accordance with the background it is placed against.

Ledia

Design: **Karsten Winkels**   Production: **Hess Form + Licht**

Two of the items of the Ledia collection are a floor-tile luminaire and a linear luminaire stripe, described as LF and LL respectively. The design and the luminous effect of these products are based on LED technology. The floor tile luminaire consists of a body built of stainless steel V4A, protected by a single-layer sheet of security glass treated with a single coat of anti-slip finish, contained by a built-in frame with a 120 mm deep anchoring system. It is merchandized in four sizes that are adaptable to most of the varieties of paving materials. The linear luminaire consists of a V4A stainless steel coffer to be built into the pavement, and a strip of single-layer security glass with a diffuser side underneath. It is available in different lengths, from 240mm to 910mm. Both luminaires can be supplied in the red, green, blue, amber and white. These fixtures can resist a weight of rubber-wheeled vehicles of up to 1.5 tonnes and require no maintenance.

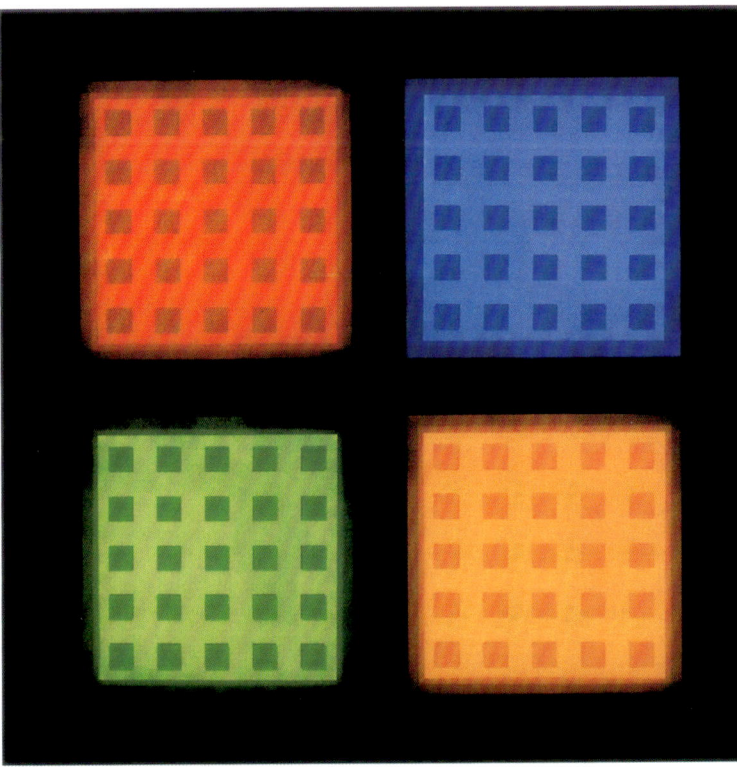

Lights

Vía Láctea

Design: **Enric Batlle-Joan Roig**   Production: **Santa & Cole**

Conceived as a way of drawing lines of light in the sky, Vía Láctea was the first and most remarkable project to use fluorescent lighting in an urban environment. The column is made from 150 x 100 mm base hot-dip galvanized structural steel sections, with a 100 x 50 mm lamppost and inspection door. The screen, which can be single for one lamp or double for two, is rectangular in section and made from the same material. It uses totally weatherproof standard HF-265 lamps for 58W fluorescent tubes, readily available on the market. The column is fixed to the ground through a reinforced concrete cube, made on site 22 cm below the level of the paving, and anchoring bolts.

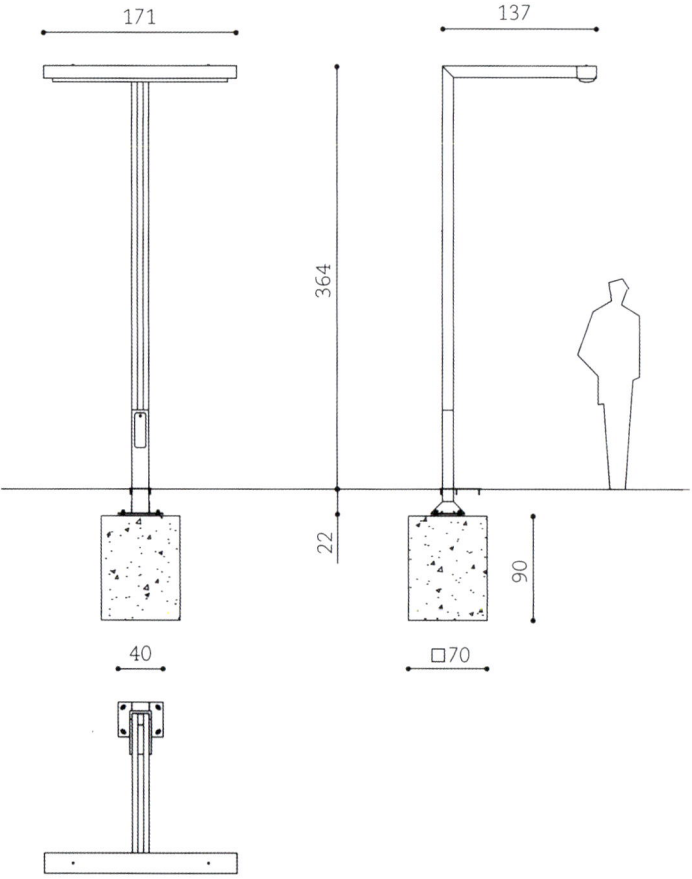

Golf

Design: **Oriol Guimerà**   Production: **Santa & Cole**

This dynamic street lamp, formed by three separate lamps at different heights, combines elements of 1970s classic street lamps with modern materials. The column is made of hot galvanized steel and consists of a base tube that is 160 mm in diameter and 90 cm in height and three upper tubes 50 mm in diameter and of varying lengths. The lamps are made of cast aluminium, sandblasted and powder painted in silver grey. The bottom cover, which supports a flat glaze-finished polycarbonate diffuser, is in sandblasted stainless steel. It uses compact fluorescence lights of 32 or 42 W. The column is fastened using a concrete cube, made on site, and anchoring bolts.

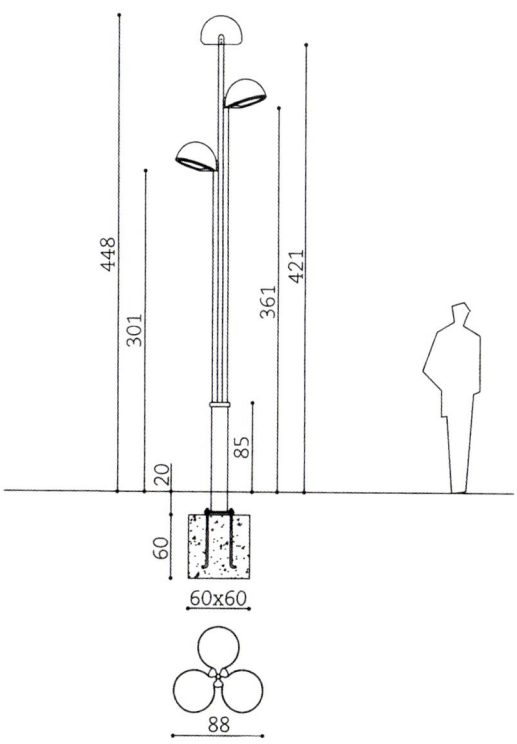

Lights   **167**

Solar Mallee Trees

Design and production: **Street and Park Furniture**

The design of these solar trees arose as part of an environmental impact research carried out by the government of Australia to demonstrate its serious involvement in the search for alternative sources of energy. The shape that inspired the project is that of the Mallee, a type of dwarf Eucalyptus tree with various ramifications, native to the environment of southern Australia. The trees were constructed in collaboration with the Street and Park Furniture Studio, who proceeded with the production and installation of the items at the Adelaide Festival Centre. The design displays a new form of dome shaped solar panels, an innovation introduced by the design studio. The dome-shaped oval panels form the treetops and provide the energy to feed the system of programmable LEDs that illuminate the Festival grounds at night. The Solar Mallee Trees include a sound system that is activated by infrared sensors and emits the soundtracks recorded in Adelaide's first "Solar Schools". The materials used to build the metallic parts of the "Mallees" are iron and aluminium; the dome-shaped photovoltaic solar panels are made of laminated fibreglass; the LEDs are controlled electronically, and emit light in the frequencies green, blue and red.

Nanit

Design: **Ramón Úbeda & Otto Canalda** Production: **Metalarte**

Nanit has been described by its designers as "a system of exterior illumination and beacon signalling". It consists of a series of fixtures to be installed outdoors, suspended from a fixture or supported on a lamppost. The support pillars of the standard lamps are made of extruded aluminium and the diffuser is made of rotomolded translucent polyethylene, lacquered white or silver. The maintenance locker is located toward the base of the pillar. The system is supplied in heights of 3.34 and 2.26 meters, with an extra-large base shoe, or in heights of 3.00 and 2.26 meters with the normal base shoe. The pillar is finished in white or in silver lacquer. The suspended lamps are also made of rotomolded translucent polyethylene, lacquered white or silver.

Limits

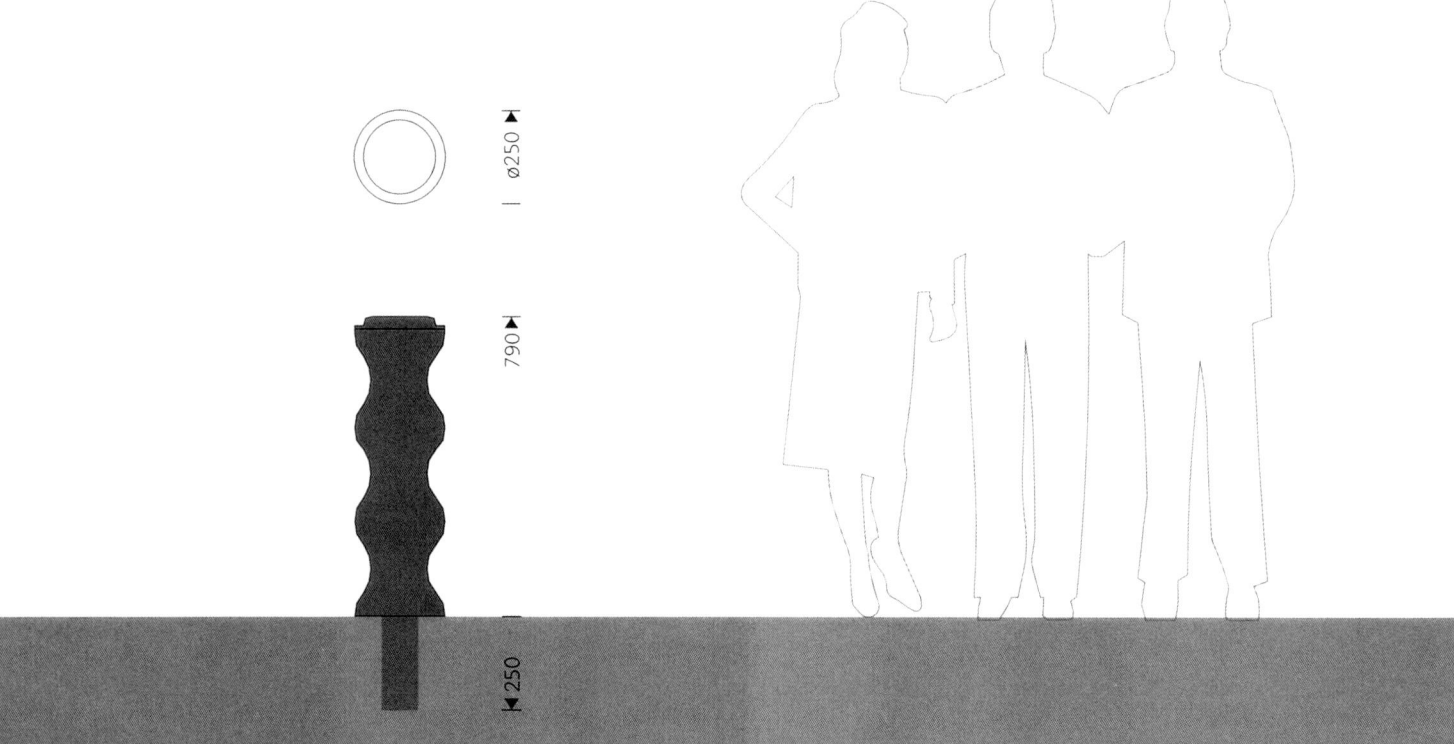

Imawa

Design: **Urbanica**   Production: **Concept Urbain**

The Imawa series consists of a single post and a barrier composed of two joined posts. The posts are 110 or 120 cm in height and are made of profiled aluminium with a factory-machined head in cast aluminium. The post is finished in powder-coated RAL colours.

Nastra

Design: **Outsign**   Production: **Concept Urbain**

The Nastra bollard has a height of 85 cm and is made of ductile cast iron. The finish is shot-blasted and metallised with oven-hardened polyurethane paint in RAL colours.

Sagrera

Design: **Josep Muxart**   Production: **Escofet**

Sagrera is a modular fence designed to limit areas in the manner of a palisade. It is made of cast reinforced concrete and has no particular surface finish. Installation involves planting the modules in a running foundation trench of 60x60cm, and then filled in with "in-situ" concrete in the customary way. Reflected light gives the items a variety of tones that play with the folds on the vertical posts, which produces an effect of light and of movement.

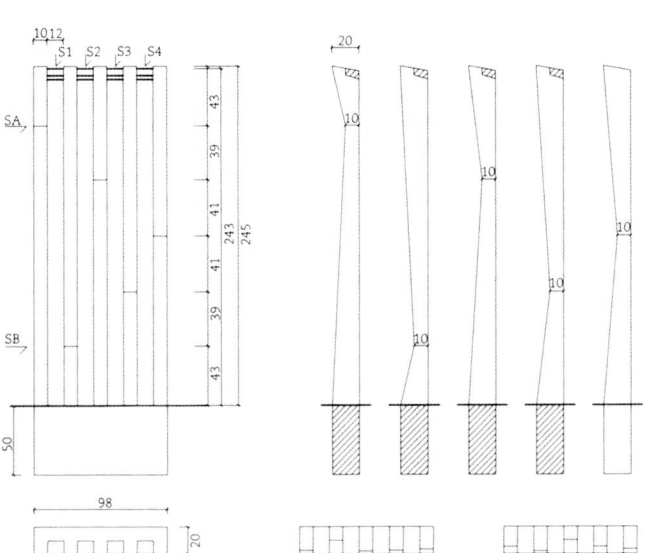

Limits

Lineapalo - Linealuce

Design: **Bernhard Winkler**   Production: **Euroform**

Lineapalo has been designed by Bernhard Winkler, professor at Munich's University of Technology. This pair of objects constitutes a simple bollard, while Linealuce follows the same design but incorporates a light.

Rondo

Design: **Bernhard Winkler**   Production: **Euroform**

The universal frame has a classic, aesthetic shape although it is contemporary in design. With a height of 89 cm, it is ideal for fencing areas off. Rondo is also well suited for parking bicycles.

Neobarcino

Design: **Joaquim Carandell**   Production: **Fundició Dúctil Benito**

Neobarcino is a pylon for the outlining of pedestrian areas. The items are made of cast iron which gives them a character of strength and resistance. Their unusual and organic form will adapt well to any urban environment.

Limits

Campus
Design and production: **GH form**

The Campus linear drainage is produced in untreated cast iron. Its dimensions are 260 × 500; it is also available in the width 140 mm. The drainage can resist weight from heavy vehicles. The standard production of this drainage unit is intended for straight lines, but can also be provided to fit the given radius of a particular project.

The design has "ears" that cover the edges of the subjacent water channel so that the drainages can be placed against surrounding pavements and in continuation of each other.

Cilíndrico
Design: **A. Viaplana / H. Piñón**   Production: **Escofet**

This modular item has been conceived to outline flowerbeds and gardens or to separate pedestrian pathways from a road. The 4 cm separation produced between the cylinders allows for the escape of excess storm-water not absorbed into the ground.

176   Limits

Onda
Design: **Atelier Mendini** Production: **Ghisamesteri**

Bollard in cast iron, UNI EN 1561, consisting of a decorative column realized in a single cast piece supplied with an anchoring extension to be secured to the ground; cast-iron cover fixed to the body by stainless steel dowels.

Marioska
Design: **Área de diseño Gitma/ Chantilly design** Production: **Gitma**

The organic looking Marioska pylon consists of a cut-out of carbon steel, lacquered in different colours according to the use it will be put to. Its shape changes the conventional appearance of a dissuasive urban item and offers the stroller a new way to visually enjoy the street. It comes prepared for two alternative types of anchoring: a base to be embedded in the concrete or with a folded plate foot, for fixing to the ground with bolts.

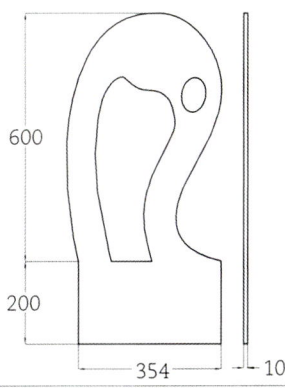

Limits 177

Petrelli

Design: **Área de diseño Gitma/ Chantilly design**   Production: **Gitma**

Petrelli is a simple dissuasive pylon; despite its straightforward and static nature, it manages to look alive. It can be placed in a row, obliquely or in a parallel distribution. It consists of a cutout sheet of Corten steel. There are two systems of fixing it to the ground, by embedding it into the pavement or with a folded foot to be bolted down.

Consentido
Design: **Área de diseño Gitma/ Chantilly design**   Production: **Gitma**

The Consentido border post offers a new variable form concept for these items, which project there interesting shadows upon the pavement and are intended as border beacons that actually indicate the direction of traffic flow. It consists of a cut out sheet of carbon steel lacquered in different colours according to the function to be fulfilled.

Limits

Street Jewellery

Design: **Damien Regamey** Presented at: **MUDAC** Curator: **Francisco Torres**

With an internationally renowned college of art (ECAL) and a design museum (mudac) that has nothing to envy museums in capital cities, design is doing well in Lausanne. It is therefore logical that a young design scene should be emerging there. It is called INOUT and defines itself as a promotional platform bringing young creative people together by means of exhibitions, publications and productions. Benefiting from a carte blanche exhibition space at the mudac and the support of sundry partners and sponsors, INOUT is burgeoning under ideal conditions. Thus, 12 young designers have implemented the museum's spaces with the project of (re) designing Lausanne's urban landscape. Experiencing the city through the eyes of these designers is equivalent to sitting down in the company of a tree, setting up one's work space in a public park, opening one's shutters onto a fruit and vegetable market, washing in a fountain, having one's path blocked by a bead necklace… so many propositions, and plenty of others that are just as unusual but imbued with a realism that is inherent to the product's design. The Street Jewellery project by Damien Regamey proposes a king-size jewellery to replace the traditional limits in the form of chains. Bling-Bling and BEAD NECKLACE are two examples.

Propeller
Design: **Greg Healey**   Production: **Street and Park Furniture**

Greg Healey worked together with Street and Park Furniture to put this new range on the market. The impetus for this furniture was a trip through the South Australian outback. Much of the realisation of this new range of urban street furniture is Greg's response to the forms, textures and colours of this extraordinary landscape. This adaptable range of furniture is particularly suitable for use in the design of contemporary streetscapes as it draws on Australian landscape forms in a unique and innovative way. Materials: painted cast aluminium with a steel internal spigot. Heavy-duty impact malleable cast iron version also available.

Rough & Ready
Design and production: **Streetlife**

The robust beams in FSC Hardwood or in All Black (see p. 14A) are combined with a solid galvanized steel or Corten steel frame. With the standard version (BOLS) the frame extends 40 cm into the ground. The floor version (BOLV) can be easily anchored on a smooth surface or substratum. The beams are covered on the top and are mounted by means of stainless-steel theft-proof flange nuts. The bollards are also available with low-energy LED beacon lighting. Available in 2 heights: 45 and 75 cm. The diameter is about 15 x 15 cm.

Rodas
Design: **Ton Riera Ubia** Production: **mago:urban**

The Rodas pylon is a dissuasive module that is highly acceptable to municipal employees responsible for the management of public space, as it will easily adapt to all sorts of urban environments. Its size and shape are also praiseworthy. Rodas is an effective solution for the purpose of controlling or directing traffic, temporarily or permanently. Suitable for pedestrian zones, garden areas, zebra-crossings or the zoning of events in public space. These circumstances require an element like this: easy to move with a fork-lift, a crane can pick it up by the ring at the top. Once it is placed on the ground it is unbreakable and unmovable.

Tube
Design: **Roger Albero**   Production: **mago:urban**

The Tube series consists of four different models of bench and a border that goes with them. These five items permit an infinite variety of spatial options to be created, adapting easily to any situation. They can also be used individually.

The essential characteristic of this series is its hollow interior that, despite its dimensions, gives the pieces a light appearance that contrasts with the real weight of the sandblasted concrete they are made of.

Solid
Design: **Tomas Ruzicka**   Production: **mmcité**

Despite its clean geometrical shape there is no lack of gentleness, enhanced by the terrazzo effect. A traditional material in a fresh, new form. An interesting detail is the system of fixing it to the pavement. High resistance and durability. The body of the bollard is made of cast polished concrete with tiny stones in a salt and pepper finish.

Bamboo

Design: **Barbato e Garzotto**   Production: **Modo**

Fence system for public and private parks, characterised by a series of vertical elements in non-continuous succession. It consists of a 3 mm thick base in Fe 360 B steel sheet angle. The wings at the base allow for an easier fixing into a pre-existing foundation. This base has 14 round tubes of Fe 360 B steel, 50 mm Ø, to be inserted sequentially: they all follow a progressive variation of the centre-to-centre distance, while maintaining their proper role as a fence. The final height of the fence can be individualized, ranging between 1280 mm to 1780 mm. The item comes finished with a hot-dip galvanising process overlaid with a coat of thermo-hardened powder-paint for outdoors. The fence is delivered in single panels of 1.5 linear metres each, which are to be placed side by side and locked together by a tongue-and-groove system.

Pisolo

Design: **Tobia Repossi**   Production: **Modo**

Bollard in sandblasted concrete, oval shaped, treated with an anti-erosion additive, with a covering cap in stainless steel. Free standing bollard.

184   Limits

Sirio

Design: **Morandi & Citterio**   Production: **Modo**

Sirio is a program of litter bins and bollards all made of AISI 304 stainless steel plate or Corten steel. The geometrical development of the laser cut sheet steel creates a rotation effect (the object looks different, depending on the viewpoint).

Robert

Design: **Miguel Milá, Bet Figueras**   Production: **Santa & Cole**

Along their usual design lines, this garden fence by Miguel Milá and Bet Figueras bears three main features: functionality, simplicity, timelessness. Derived from the classic traditional fence we used to protect our gardens, Miguel Milá and Bet Figueras have developed a new light, long-lasting fence, noticeable for its great resistance and easy platen replacement. Made of only two elements a cylindrical support and a double-platen sheet this sober yet, at the same time, functional piece easily adapts to any garden type or city ground. Designed as a modular piece, it can be used in different ways (fits any surface) and installed in open or closed sequences. Made of laminated sheets and stainless steel rails, the Robert fence is a versatile and resistant piece that will remain in its place for an unlimited time.

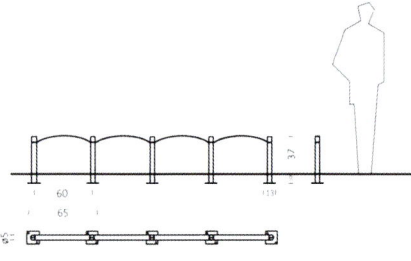

Paracelso

Design: **Adolfo Ruiz de Castañeda**   Production: **Tecam BCN**

Valla Paracelso is a railing that spans lengths of 2.13m. It consists of a square vertical post with a 135mm. The post has a series of square perforations of 20x20mm, through which light shines out. Its horizontal part is formed by an oval handrail, and four cylindrical tubes below, with a 220mm gap between them; it can also be supplied to order with only two tubes. The distance of 2 m between posts can vary according to the requirements. The materials used for the item are cold galvanized carbon steel, or stainless steel, or the combination of both. It can be paint finished or semi-gloss finished if it is the stainless steel model. The fence can be leaned towards the pavement or the other way round. It is fixed with metallic M10 raw plugs. The electrical equipment consists of a compact Ip 67 lamp with an 18-watt fluorescent tube.

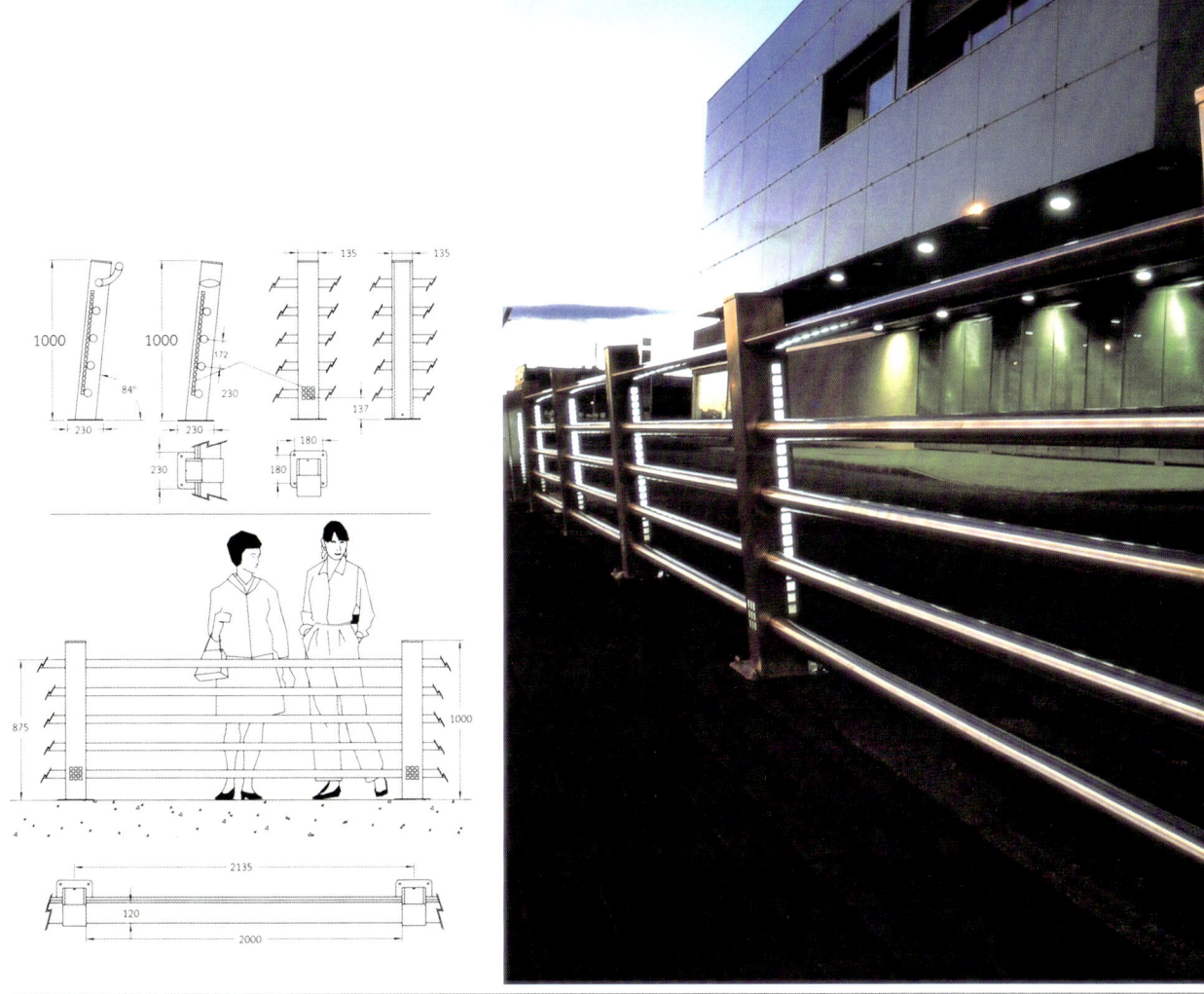

Limits

2994
Design and production: **Neri**

Bollard made in UNI EN 1563 nodular cast iron. The bollard is composed of a helical cast iron column with a rotation of 63° around its vertical axis. The base of the column has a flange. The bollard's height above ground is 95 cm.
The bollard is fitted with a cast iron fixing lug cast together with the bollard, for cementing into the foundation plinth.

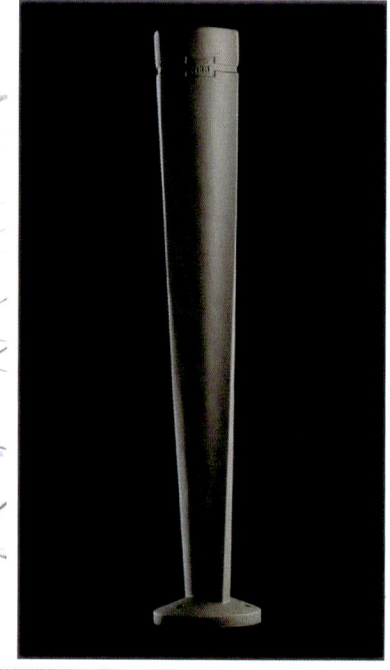

Topino
Design and production: **Geohide**

Topino is a robust box-shaped road marker designed to be used on streets and roadways, which includes a lighting system built into one of its sides. The body consists of a prefabricated white reinforced concrete element, with a sandblasted finish and smooth corners and sides, although other finishes are also possible. It measures 0.30 x 0.30 m, with a height of 0.60 m. Its total weight is 155kg.

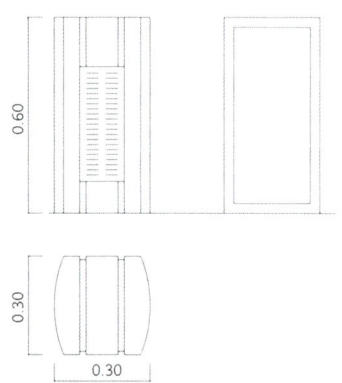

Linea
Design: **Montse Periel**   Production: **Santa & Cole**

This is a stainless steel line; it is a limit and a place to rest. In just one item, it provides a solution to the problems of co-ordinating urban space. This linear handrail acts as a barrier to pedestrians and as a place to lean on. It restricts access, marks out areas, acts as protection against falls, or subtly accommodates the public. It is a minimum formal expression and offers many functions without being a visual barrier. The single material it is manufactured in, stainless steel, provides a light appearance and is essentially hardwearing. It is very sturdy and all its parts may be easily changed. Its modular design means it can be installed in continuous sequences, according to the need.

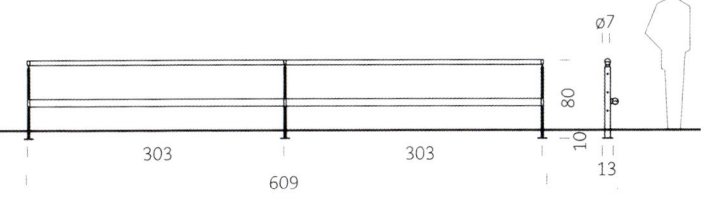

Limits

Planters

Bilbao

Design: **Josep Muxart**   Production: **Escofet**

The Bilbao series consists of a planter and a bench made of cast reinforced concrete. It has been designed to lend an organic and gentle look to the rigidity of the material. Thus, it features a mixture of curved lines and warped surfaces. The items are disquieting and seem to move discreetly. Their size makes them environmentally friendly. The concrete is sandblasted and waterproofed, which gives the material the appearance of natural stone. The pyramid shaped planter has a triangular base and the sides are slightly convex. The final shape is obtained by a circular torsion displacement of the upper plane in respective of the lower plane which generates arched surfaces. The angle of vision also influences it, creating a sequence of apparently different yet similar objects.

Dara

Design: **Josep Suriñach** Production: **Fundició Dúctil Benito**

Dara is a planter of simple and elegant lines, designed for public spaces and pedestrian areas. It is made of iron which makes it highly resistant to corrosion, resisting up to 300 hours of saline fog. The finish is in RAL colours. Dara is inspired in the traditional garden flower pots, thereby giving an intimate touch to the spaces where they are installed, especially suited for quiet corners of the city.

Planters

Conical Tree Tubs

Design and production: **Streetlife**

These Conical Tree Tubs are available in various diameters, heights and material. They are made of CorTen steel or moulded in thick-walled, insulating synthetic material.
The CorTen steel makes the lifespan of the Conical Tubs almost unlimited (they will never rust through). If required, they can be finished in a RAL color and equipped with a municipal logo.
The standard version of the Syntetic Conical Tree Tub is mat, light grey or anthracite with a granite effect. The two available heights, 110 and 150 cm, permit light-footed layouts. The volume of 1.8 to 2.3 m³ is suited for trees and shrubs of up to 5-6 m tall. The tubs contain facilities to anchor the root ball, and to ensure insulation, good drainage, and the in-flow of air.

Highlife Tree Tubs

Design and production: **Streetlife**

The Highlife tubs are part of the Highlife range, which includes benches and tree-grills. At ground level, these Highlife Tree Tubs can be easily relocated with a forklift truck. The tubs are equipped with FSC Cumaru wooden strips 25mm thick (highest durability class). They are suitable for trees of 12-14 meters. The tubs match perfectly with the Highlife benches.

Shrubtubs
Design and production: **Streetlife**

The Shrubtubs are made of 4mm CorTen steel as standard. They contain a volume of growth medium up to 5 m³ and are suitable for shrubs, multi-stemmed trees and bigger city trees depending on the substrate volume. The Treetec nursing system is optional. The CorTen tub is well detailed, functional and deliverable with special double powder coating. The Shrubtubs are easy to move and anchoring of the root ball is possible. The single sides are strengthened to prevent bending by soil pressure. The bottom has special holes for water drainage and oxygen supply. The walls should be equipped with thick-walled geo-textile for good insulation and oxygen distribution.

Urbe

Design: **Joan Gaspar**   Production: **DAE**

The Urbe planter is made of 6 mm thick sheet iron, with a powder coated finish that makes it resistant to corrosion. Moreover, it comes powder-coated with lead-free polyester, which gives greater strength to the exterior finish, with a black wrought-iron effect. The boards are of tropical wood, treated with tinted transparent oil, to enhance the wood's natural beauty and give it maximum protection.

Calipso

Design and production: **Modo**

The Calipso planter series includes three planters, one big, one small and one wall mounted. They are made of 5 mm moulded steel sheet. The sinuous line at the base was designed to give an impression of continuity if the flower boxes are placed in succession or combined with each other. This flower box proposes a form of decoration introduced discretely in the urban context.

Iona

Design: **Tobia Repossi**   Production: **Modo**

A circular planter made of two rounded sheets of Corten steel joined by means of a ring-shaped section in AISI 304 stainless steel. There is an "eight-shaped" hole for the flowers in the centre of the planter. When supplied, the planter doesn't have an oxidised surface, which will appear after a period of natural "aging", due to exposure to atmospheric agents. Iona is available both in the convex and concave version. It is a freestanding planter.

Lineafiorne

Design: **Bernhard Winkler**   Production: **Euroform**

The Lineafiorne planter designed by professor Bernhard Winkler has a reservoir made of 2 mm metal sheet, hot-dip galvanized and powder coated with RAL colours. This reservoir is covered with untreated hardwood slats 30x59 mm, and finishing wooden slats 60x65 mm. It is a freestanding planter that does not require an anchoring system.

Lolium

Design: **Área de diseño Gitma / Chantilly design**   Production: **Gitma**

This is a modular planter made of Corten steel. It features a sharp slope that shows the flowers it contains to their best advantage. This is achieved by means of a straight piece placed at a 60° angle that permits a variety of combinations. It requires no special fixing to the ground as the weight of the earth prevents it from moving.

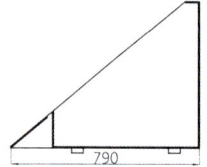
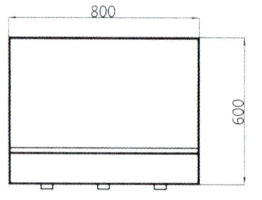

Marc
Design and production: **ONN Outside**

The Marc planter features an upper frame to enhance its decorative function and show off its contents to their best advantage. As it is available in two sizes and finishes, this planter covers the requirements of different plant types and is suitable for use in any urban environment. The structure is of Corten steel or zinc plated steel with a two-tone epoxy finish. The bolts and fixtures come in stainless steel.

Planters 199

BdLove Planter
Design: **Ross Lovegrove**   Production: **Bd**

The BdLove Planter is made of medium density polyethylene, roto-molded and pigmented, containing five stainless steel reinforcement bars. It also has a self-watering system that consists of a circular disk with its capillary absorption cords, a water level indicator, a water supply pipe and three drainage bags. Non-slip EPDM rubber feet allow the water to drain correctly underneath the BdLove Planter, which is attached to the floor by means of stainless steel bolts. The planter comes in the following colours: fluorescent red (only for indoor use), beige, white, blue, green and sandstone (sand coloured granite) and millstone (dark coloured granite). It can also be custom made in special colours. The units are stackable and each unit measures 1.356 x 1.000 x 483 mm and weighs approximately 27 kg. The planter can be weighed down with water or clean sand, for extra stability.

Sputnik
Design and production: **Colomer**

This very robust circular container for plants has a clearly sculptural outline. It is made of grey coloured cast iron according to the quality norms UNE-EN-1561. The joints are sealed with zinc phosphate primer. The legs are available in a range of different finishes and materials. The entire item has received an anti-corrosive treatment and comes with a wrought iron finish. The various component parts are held together with zincified steel screws. The planter has the noteworthy capacity of 332 litres and its weight amounts to approximately 260 Kg. The only installation required is to place it on the ground in the desired location.

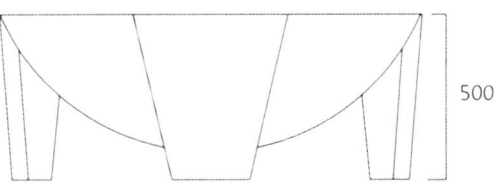

Planters

Tree Grids

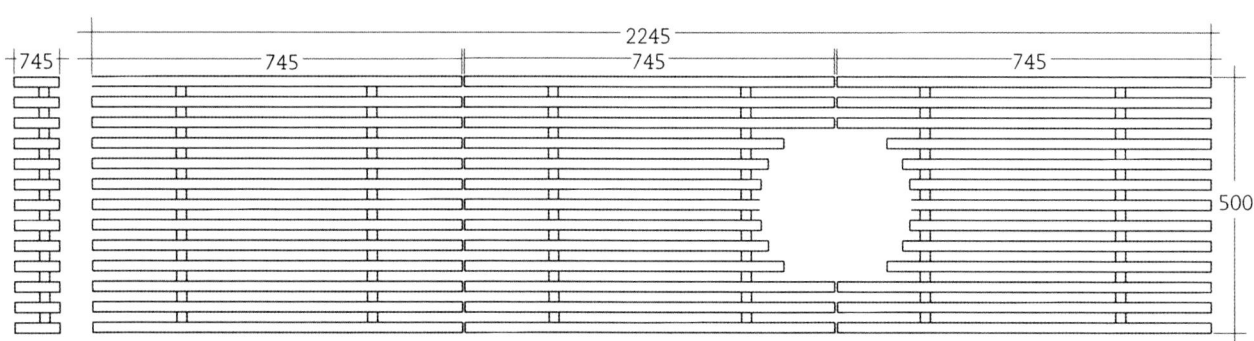

Rámla

Design: **Pere Cabrera & Jaume Artigues**   Production: **Escofet**

The name of this tree grid comes from the Arab term "ramla", meaning gravel or sand-pit, which later derived into the popular Latin term "Rambla", indicative of a torrent or dry river bed that only carries water after heavy rainfall. This item is designed for paved areas in the city, to distinguish or define the area the trees are planted in, and allows storm-water to flow in or out freely. In the Mediterranean area, the presence of torrential rains enables tree-holes to be virtually self-cleaning. There are various different formats and accessories that allow a versatile adaptation of this item to existing trees: the tree grid can be enlarged or contracted as needed.

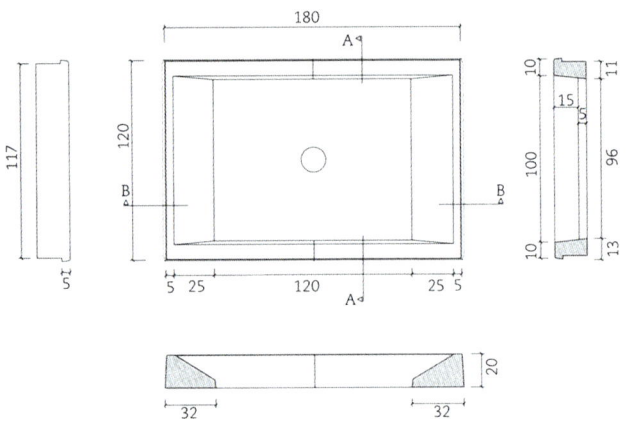

Sombra

Design and production: **Durban Studio**

The design of the Sombra tree grid is a visual analogy of the environment in which it will be installed, as the criss-cross pattern of the grid repeats the shadows of the branches on the pavement. It consists of two parts that fit together, surrounding the base of the tree-trunk.

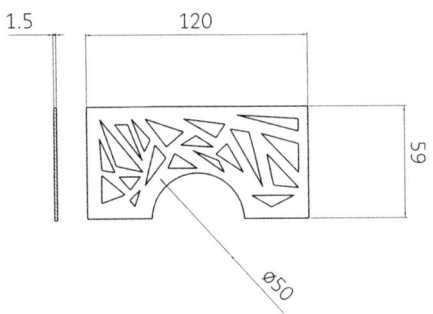

Beiramar
Design: **Guillermo Vázquez Consuegra**   Production: **DAE**

The Beiramar tree-grid consists of a modular piece made of cast aluminium. The perimeter is reinforced and it has radial reinforcement ribs. The object is pierced by a series of leaf-shaped drainage holes that spread outward from the central hollow. The frame around the tree-hole is made of hot-dip galvanized steel plate, screwed onto the surrounding concrete pavement.

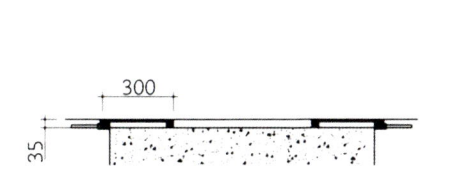
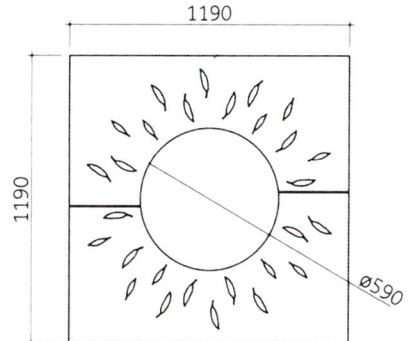

Taulat
Design: **Bernat Matorell Pena**   Production: **Fundició Dúctil Benito**

This tree-grid has a simple and functional design that permits the tree to grow without bursting the tree-grid or the pavement. The arches can be cut out according to the girth of the tree, so this item can easily adapt to any tree and remain functional during the growth of its entire life-span.

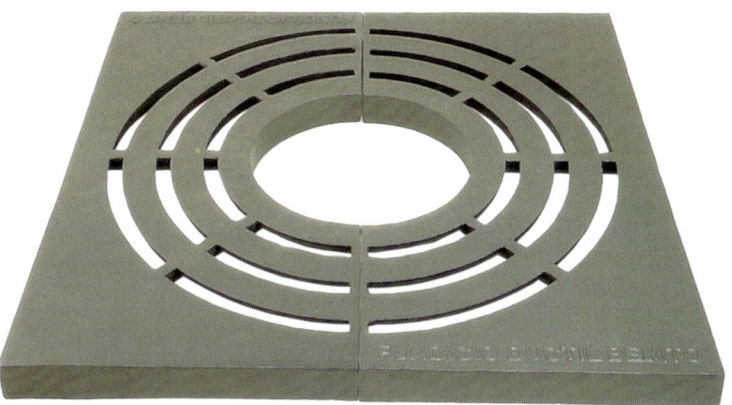

Tree Grids   205

Campus

Design: **SLA**   Production: **GH form**

Both types of tree grid come in untreated cast iron and consist of three sections. The rectangular grid has one continuous piece and two others with an identical semicircular hole that provides for flexible installation possibilities in relation to the surrounding pavement and asymmetrical placement around the tree. The grids are provided with a steel frame.

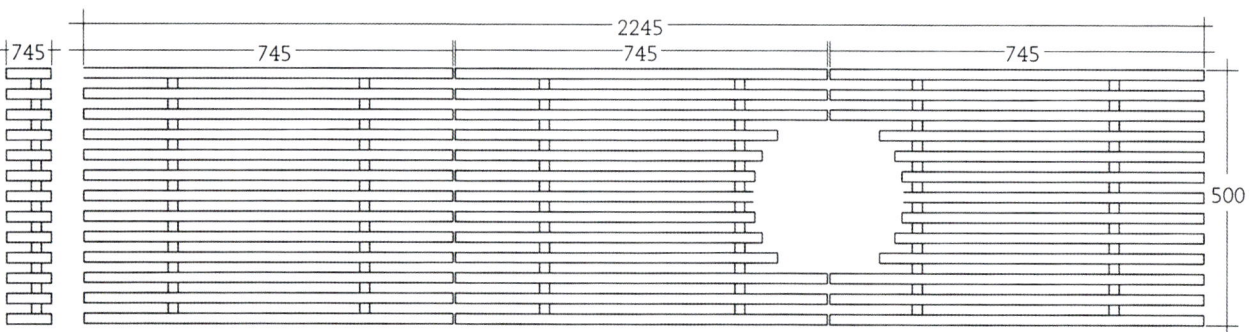

Tree Isles

Design and production: **Streetlife**

The circular Tree Isles come in three sizes with diameters of 3, 4 or 5 meters. The tree isles consist of laser-cut CorTen steel sections whose edges can be set into a perimeter of FSC hardwood that functions as a seat. Tapered filling wedges between the beams seal the seating perimeter. The tree isles are completely freestanding on the concrete deck or other material. Streetlife also produces large Tree Isles in oval and free forms, thus converting hard urban locations into sympathetic sitting and resting places. A good reference point is the Bastiaansplein in Delft, with its three oval tree islands (9 × 4 m) with large pagoda trees above a car park.

CorTen Tree Grids
Design and production: **Streetlife**

The thick-walled CorTen steel Tree Grids are equipped with a random dot pattern. They are available in a round and a square variant, both with a tree hole of Ø 35 cm. Made-to-measure items can be customized in terms of dimensions and graphic pattern.

  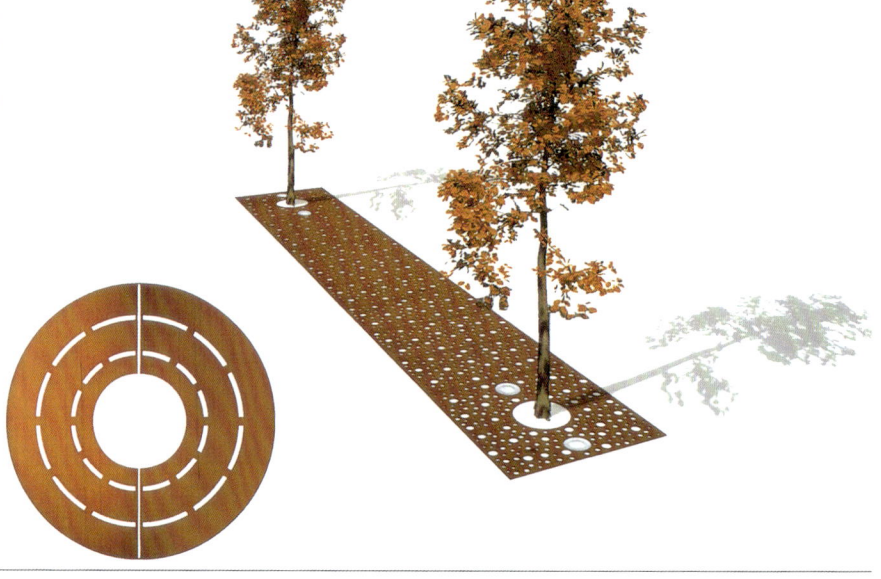

Rough and Ready Big Green Benches
Design and production: **Streetlife**

The Big Green Benches are made of CorTen steel with an integrated FSC hardwood bench. In the Rough & Ready version, this sitting rim is splendidly detailed, with the wooden beams inlaid. These benches can be filled with shrubs, bamboo or climbers. The benches are extremely solid but remain movable. It is possible to supply benches with an open bottom (OB version) so that shrubs or trees can take root in the open ground. They are also extremely suitable for bamboo or climbers for façade greenery.

Helix

Design: **Greg Healey** Production: **Street and Park Furniture**

The impetus for this piece was a trip through South Australia's outback. Greg has responded to the forms, textures and colours of this extraordinary landscape in the realisation of much of this new range of urban street furniture. This adaptable range of furniture is particularly suitable for the use in contemporary streetscape design as it draws on Australian landscape forms in a unique and innovative way.

Cortengo

Design and production: **Velopa**

The Cortengo fits in perfectly with the newest trends in architecture due to its design and choice of materials. The particular construction of the square openings creates an abstract image of root development. Corten steel undergoes a natural oxidation process that reinforces its special image. A unique and unpredictable play of colours is created that makes every Cortengo a colour experience of its own.

Cap i Cua

Design: **Juan Carlos Bolaños** Production: **mago:urban**

The Cap I Cua tree-grid consists of two identical elements made of structural concrete with a resistance of 400 Kg/cm2. The item is entirely self-supporting, resting directly upon the layer of concrete underneath the paving material, thereby making it very easy to install. Moreover, its dimensions are in accord with the various formats of industrially produced paving materials (20x20, 30x20, 30x30, 40x40, 60x40). It is very resistant: with a thickness of 8cm, it is guaranteed to support even vehicles with a load of up to 900 Kg per wheel.

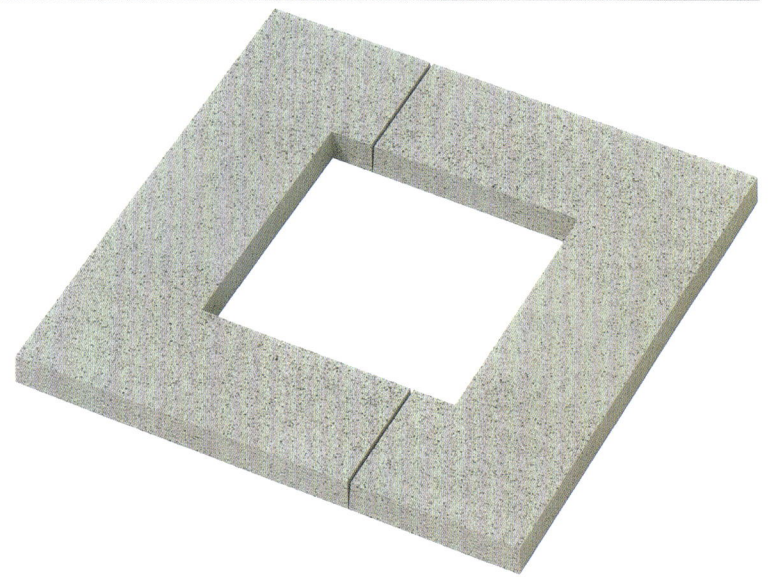

210 Tree Grids

TV

Design: **Ernest Perera**   Production: **mago:urban**

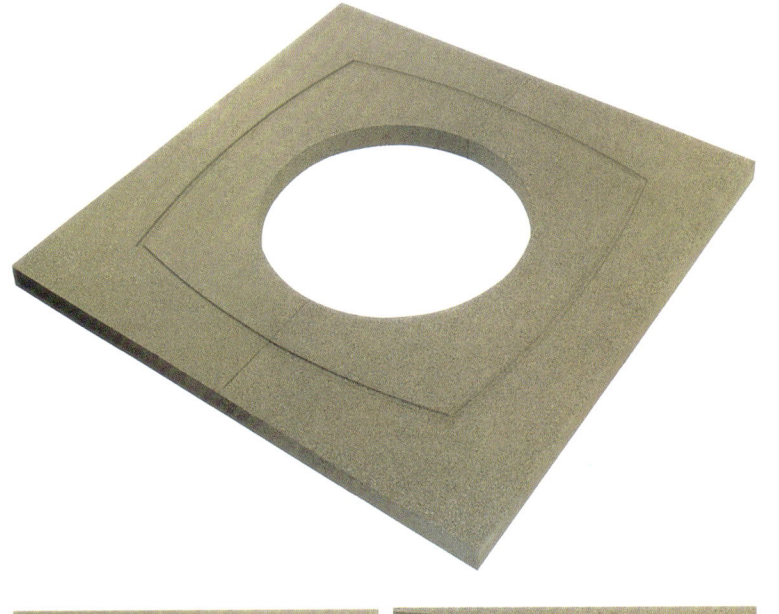

This tree-grid consists of two identical parts made of structural concrete that has a shearing stress resistance of 83 kg per cm2 and withstands a compression stress of 350 Kg per cm2. It is easy to install as it can be rested directly on the concrete slab beneath the pavement. It has a series of runnels all around the perimeter to enable the entrance or exit of water. Its dimensions are such as to make it compatible with a range of industrially produced paving materials (20x20, 30x20, 30x30, 40x40, and 60x40). Its thickness of 8cm is guaranteed to support even vehicles with a load of up to 900 Kg per wheel.

Tree's Stool

Design: **Émanuelle Jaques**   Presented at: **MUDAC**   Curator: **Francisco Torres**

Tree's Stool is part of the INOUT artistic program for redesigned public spaces in Lausanne, sponsored by the mudac design museum. As a mutation between a protective barrier for a tree and a seat, this piece of furniture provides an opportunity for passers-by to rest a while.

Bicycle Racks

On

Design: **Ramos / Bassols**   Production: **Alis**

On performs as both bicycle rack and bollard. It is made of 5mm thick steel plate. Its circular base (220 mm diameter) permits it to be anchored to the pavement by a series of holes made for that purpose. The vertical part of the item is made of 15mm thick steel plate and measures 560x222 mm. The hole at the top has a diameter of 185mm that allows chains and rods to be passed through to lock bicycles to it. The various structural elements are arc-soldered together. The finish is galvanized by immersion in zinc, which is then dust sprayed in epoxy gauge 150 micros, giving it a smooth and even look. Colour can be customized.

Trian

Design: **Runge Designteam** Production: **Runge**

Trian displays optically generated tension and dynamics. In addition to its characteristically dynamic shape, Trian is a safe cycle rack. The round tube made of electropolished stainless steel has a diameter of 60 mm. It is asymmetrically welded with mitre cuts of 45° and 29 °.

Sammy

Design: **Oriol Guimerà**   Production: **Santa & Cole**

A modular piece, at once bollard, containment rail, bicycle rack and potentially even a seat, made of cast iron and extruded aluminium. A good solution for defining functional boundaries of urban space and demarcating the city. The support is made of nodular cast iron with anti-rust protection and painted in black. The base is inserted 15 cm into the pavement into a previously drilled hole and filled with epoxy resin or fast-drying cement or such. No maintenance is required. Weight: 34 Kg.

Flo

Design: **Brian Cane**   Production: **Landscape Forms**

Flo is a horizontal fluid spiral form of handsome and robust stainless steel tubing designed by Brian Kane to make a row of bicycles into a contribution to urban space while holding them securely and safely. All bike racks by Landscape Forms meet APBP (Association of Pedestrian and Bicycle Professionals) recommendations for supporting bicycles at two points and locking in at least one. All the racks are strong, durable, weather-resistant and tested to meet the quality standards requested by Landscape Forms, which led to awarding them the LEED credit for the encouragement of environment-friendly bicycle use.

Táctil

Design: **Antonio de Marco**   Production: **Santa & Cole**

A new kind of bike rack, a cheerful, uninhibited and very sturdy bollard, designed to cover the ever-increasing demand of parking space for city bicycles. Made of one single piece of sandblasted Corten steel sheet, silhouetted using an oxycutting process. The piece is delivered fully assembled and with anchoring instructions. The shaft is fixed into the ground with two bolts through the same piece, which go into holes previously filled with epoxy resin, or fast-drying cement or such. Maintenance is not required. Weight: 20 Kg.

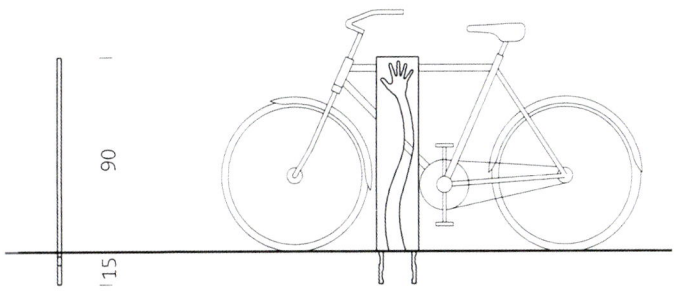

Bicycle Racks

Contínuo - Triângulo

Design: **Pedro Pereira - Francisco Providência**   Production: **Larus**

Contínuo - Triângulo are two models of bicycle rack made of galvanized and painted steel. They are held to the ground by means of their own specific anchoring system. The Contínuo model is fixed by means of a perforated steel base-rod. The Triângulo model is designed to hold the bicycle directly by the frame.

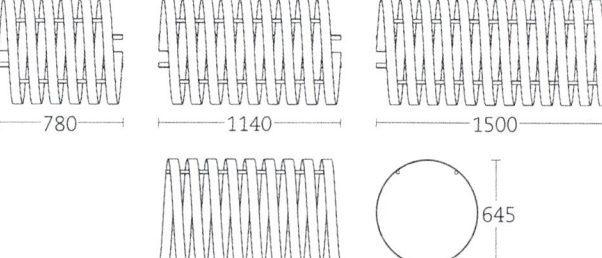

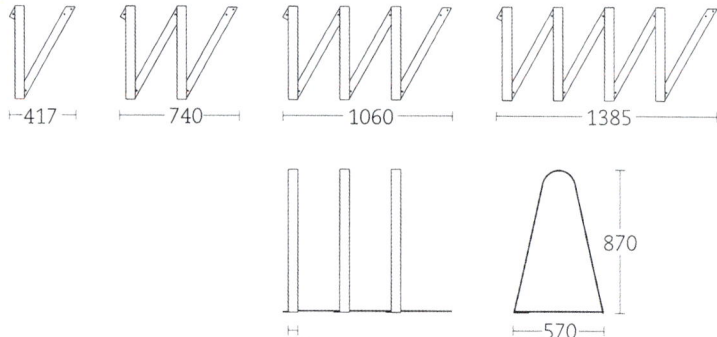

Bicycle Racks

Key

Design: **Lagranja**   Production: **Santa & Cole**

The design of this bicycle rack is based on agreeable, simple and dynamic lines. The idea was to revitalize a rather neglected urban element. The structure is of high-density polyurethane and the base is of cast aluminium. It comes in two standard colours, anthracite gray and red.

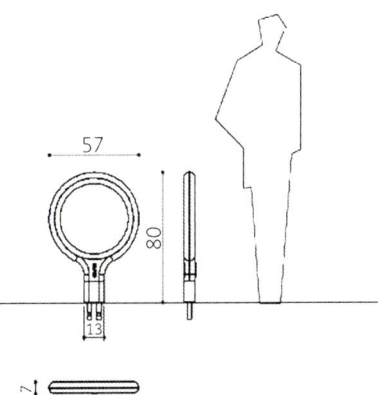

Bicycle Racks

Ezzo

Design: **Tomas Ruzicka** Production: **mmcité**

The natural beauty of traditional terrazzo technology enhances the purist form of the Ezzo line of bicycle stands. The separate units can be assembled into endless lines. These bicycle stands are made of cast polished concrete containing tiny stones in a granite-like salt and pepper texture. The units are held together by a galvanized steel bar. The rack's solidity and weight allow it to be installed as a freestanding installation on the ground.

Bicycle Racks

Meandre
Design and production: **Mmcite**

This steel frame holds a sturdy ripple of rubber belt sinuously curved into a meandering shape for slipping up to five bicycles into the grooves resulting on either side. The horizontal bar at the top provides the element to which the bicycle can be chained and locked.

127

Design and production: **mago:urban**

The bike station 127, conceived and manufactured by mago:urban, features the possibility of parking three bikes at the same time. It also includes a ring that can be used with any kind of bike lock. The ends have been designed to allow more station 127s to be connected, as and when space for more bikes is needed.

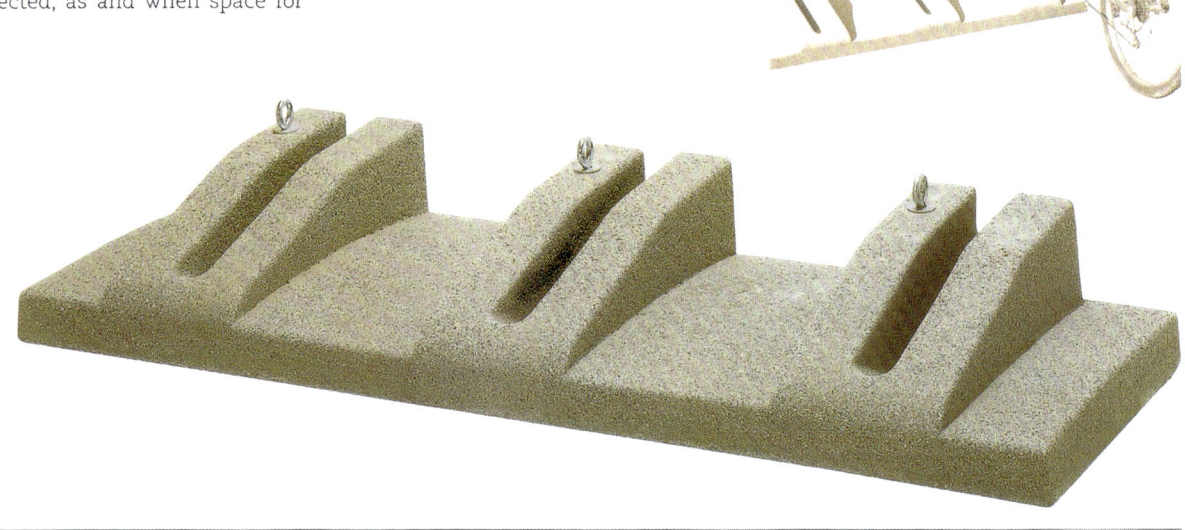

Rough&Ready

Design and production: **Streetlife**

The Rough & Ready bicycle parking rack is a tough supporting and durable piece of urban furniture. The untreated and sustainable FSC hardwood (FSC controlled wood means it did not originate in protected zones or endangered species). The stand is designed not to damage the bicycles and creates a new and idiosyncratic image. For the first time, the street can feature a natural-looking, user-friendly yet robust stand. The hole in the steel leg is for securing the bikes to the stand with a chain lock.

VD 003

Design: **Adrien Rovero**   Presented at: **MUDAC**   Curator: **Francisco Torres**

VD 003, by Adrien Rovero, is part of the INOUT artistic program for redesigning public spaces in Lausanne, Switzerland, sponsored by the MUDAC design museum. This bike rack occupies the space of a single car parking space and offers parking spaces for up to six bicycles. As a bicycle rack, it is both functional and addresses a critical statement to car traffic.

Shelters

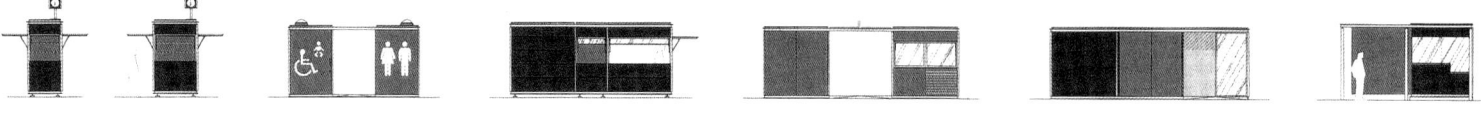

Andromeda

Design: **Studio Rota & Partners**   Production: **Modo**

The contemporary shape of this bus shelter creates a charming aesthetic impact. The inner structure is made of Fe 360 B steel, covered with panels of decorative laminated plastic suitable for outdoor use. The bus stop shelter is available with or without plate glass sides, with wooden seats and a lighting system.

Heritage

Design and production: **JCDecaux**

Heritage is a shelter with a capacity for 15 persons, six of them seated. The structure is supported by four cast iron posts. The roof measures 5 m2 and the information box comes as a separate accessory. The glazing consists of three separate sheets of glass at the back and one at the side. The materials chosen for this item provide maximum durability and resistance, and its design is vandal-proof.

Refugio Verde

Design: **Diana Cabeza, Martín Wolfson**   Production: **Estudio Cabeza**

Refugio Verde shelters a bus stop in Puerto Madero (Buenos Aires). The roof is of laminated security glass with an inner layer of green PVE, designed to act as a green filter onto which the leaves seem to have dropped after a strong gust of wind, creating a sort of green canopy. In the summertime the users find shade here, and for a while their skin is tinted a beautiful shade of green. The structure is made of steel, protected from corrosion with polyurethane paint.

Shelters

Edge

Design: **David Karásek - Radek Hegmon** Production: **mmcité**

The Edge bicycle shelter has a radial dynamic shape that combines a robust supporting structure with fine glass walls held up by sturdy stainless steel holders. The bicycle wheels are held in grooves that pierce the building's distinctly slanting plastic rear wall, the shelter's most characteristic feature. The frames can be chain locked to the tubular steel bars that separate the bicycles. The steel structure is hot dip galvanized or powder coated in any RAL colour. The roof and sidewalls are made of tempered glass. Optionally it can be provided with a corrugated roof of zincified sheet-steel.

Nimbus

Design: **David Karásek - Radek Hegmon**   Production: **mmcité**

The stable and time-tested structure of this range of all-purpose shelters enjoys a visually light appearance. Its smartly supported arched roof seems to hover over the glass walls and the slender steel pillars. The shelter is available in a wide selection of dimensions. The galvanized steel structure is painted in a standard shade. The rear and side walls are made of tempered glass, the roof consists of twin-wall polycarbonate. Rain water drains through the pillars. The shelter has a solid wooden seat treated for outdoor use. A timetable holder is optional.

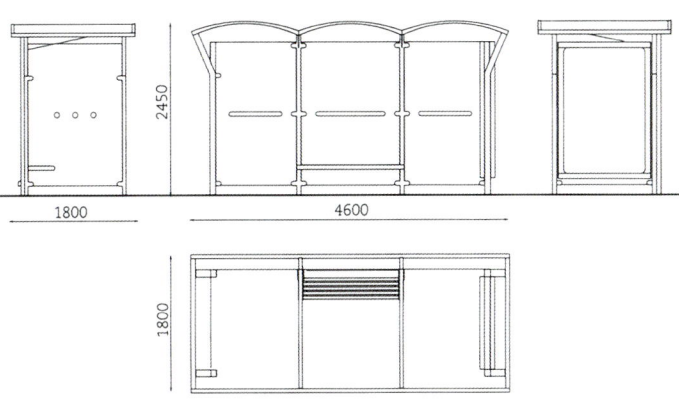

Shelters   **229**

Skandum

Design: **David Karásek - Radek Hegmon**   Production: **mmcité**

A simple but elegant structure that fulfils all the requirements on cost-effective terms. The heat-galvanised steel structure comes painted in a standard shade. The rear and side walls are of toughened glass. The roof is of chamber polycarbonate. Rainwater drains through the pillars. The shelter has a solid wooden seat treated for outdoor use. A timetable holder is optional.

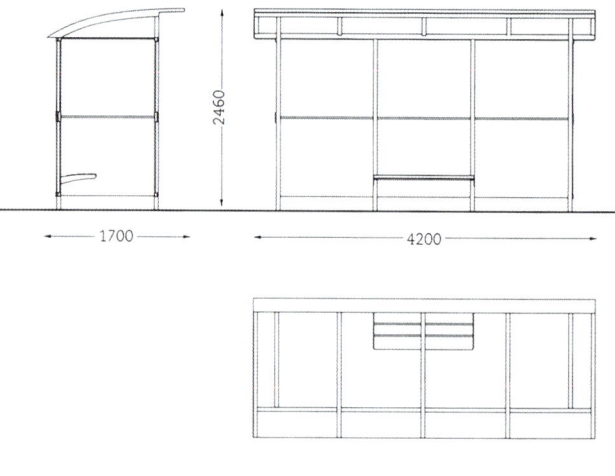

Kaleidoscope
Design and production: **Landscape Forms**

Kaleidoscope is a series of parts designed to create a sense of place. The basic structure consists of a post and a canopy. These can work on their own or with the optional addition of seating and lighting to make handsome gateways, transit stops on city streets, valet parking stations, bike rack shelters, break areas on corporate campuses and covered seating in malls, parks and zoos. Kaleidoscope is modular and has replaceable parts. It is assembled, not constructed, and can be removed and relocated. It meets demanding wind force and snow load testing.

Smart Shelter is an outdoor transit shelter that integrates the Kaleidoscope canopy and structure by Landscape Forms with solar powered lighting and advanced light-emitting diode (LED) technology.

Smart Shelter was developed in response to customer requests for an illuminated transportation shelter requiring no external power and reliably meeting the tough demands of urban environments, a rugged, durable, cost-effective and low maintenance solution for cities and transit agencies that enhances safety, security and service.

Habana

Design: **Màrius Quintana**   Production: **Microarquitectura**

Microarquitectura has installed the Habana kiosks in parks and development projects in Barcelona and neighbouring towns such as Terrassa, Sabadell, Sant Adrià, Badalona, Ripollet, Santa Coloma, El Prat de Llobregat, Sant Boi, Vilanova i la Geltrú, Cambrils, el Vendrell, Lloret de Mar and all the way to Valencia, Cáceres and Ibiza. The kiosks have been approved by the Barcelona Town Hall, Barcelona Institute of Parks and Gardens, Collserola Consortium, Barcelona Metropolitan District, Cambrils Board of Tourism, The Sagrada Família Board of trustees, and others. Microarquitectura regularly carries out signaling projects or sculptural installations. As for the urban furniture they develop, the ADA series is a relevant example of beach facilities (shower, foot shower, fountain and litterbin) or the AJC series of streetlamps, benches and litterbins. The enterprise is experienced in working jointly with institutions in a context of research, to provide items of tested quality in which performance and sustainability meet.

Designed to fulfil the requirements of seaside, park or urban environments, Habana is a module that can be installed with or without the pergola. The two items, module and pergola, have been separately designed to enhance their adaptability to any situation. The pergola can be extended indefinitely or even be installed on its own; the sunscreen roof of wooden slats can become an awning if replaced by weatherproof glass. The underlying idea is to implement the maximum respect for urban locations through unobtrusive, timeless design. The unconcealed metal structure can accommodate a variety of wall panels, fixed or sliding, opaque or clear. When closed it becomes a vandal-discouraging compact volume, which unfolds into a welcoming open space during working hours. This tightly measured object allows transportation and assembly to be executed with no complications beyond connecting it to the power and supply line. Built according to accessibility requirements for the disabled, it can house a cafeteria, a vending stand, a first-aid centre or a dressing room.

Company index

Alis

www.alis.es
Tel: +34 93-727-61-72
Fax: +34 93-726-55-26
E-mail: alis@alis.es

12, 13, 22, 23, 134, 135, 214

bd

www.bdbarcelona.com
Tel: +34 93-457-00-52
Fax: +34 93-207-36-97
E-mail: export@bdbarcelona.com

94-95, 200

Colomer
Altimira, 5-7 Polígono Industrial Santiga
Apdo. Correos 99
08210 Barberà del Vallès
Barcelona
Spain

www.colomer-es.com
Tel: +34 93-719-08-52
Fax: +34 93-718-78-88
E-mail: info@colomer-es.com

201

Concept Urbain
ZI Fosse Neuve
37210 Parçay-Meslay
France

www.concepturbain.fr
Tel: +33 02-47-29-07-08
Fax: +33 02-47-29-07-09
E-mail: export@colomer-es.com

24, 111, 172

DAE
Marcel·lí Gené, 16
Pol. Ind. Sant Jordi
08800 Vilanova i la Geltrú
Barcelona
Spain

www.dae.es
Tel: +34 93-814-94-00
Fax: +34 93-893-33-58
E-mail: export@dae.es

25, 89, 196, 205

Durban Studio

www.durbanstudio.com
Tel: +34 659-217-608
Fax: +34 972-574-942
E-mail: info@durbanstudio.com

26, 27, 28-29, 204

Escofet

www.escofet.com
E-mail: informacion@escofet.com

18, 19, 56-57, 58, 59, 60-61, 62, 87, 102, 103, 108, 109, 136, 173, 176, 192, 204

Estudio Cabeza
Serano 1249
Pasaje Soria 5020
C1414DEY
Buenos Aires
Argentina

www.estudiocabeza.com
Tel: +54 11-4772-6183
Fax: +54 11-4777-0811
E-mail: info@estudiocabeza.com

31, 32, 33, 85, 100, 101, 112, 128, 227

Euroform
Via Daimer 67
I-39032 Campo Tures (BZ)
Italy

www.euroform-w.it
Tel: +39 0474-678131
Fax: +39 0474-678648

30, 34, 35, 110, 174, 198

Fundició Dúctil Benito
Vía Ausetania, 11
08560 Manlleu
Barcelona
Spain

www.benito.com
Tel: +34 93-852-1000
Fax: +34 93-852-1001
E-mail: info@benito.com

35, 113, 137, 138, 175, 193, 205

Geohide
rue Saint-Martin 7
CH - 1003 Lausanne
Switzerland

www.geohide.ch
Tel: +41 21-351-50-61
Fax: +41 21-351-50-25

188

GH form
Baekgaardsvej 64
DK-41 40 Borup
Denmark

www.ghform.dk
Tel: +45 59-44-0990
Fax: +45 59-44-0440
E-mail: mail@ghform.dk

36, 37, 115, 156, 176, 206

Ghisamestieri
Via Grande 226
47032 Bertinoro [FC]
Italy

www.ghisamestieri.com
Tel: +39 0543-462-611
Fax: +39 0543-449-111

16, 17, 114, 140, 141, 164, 177

Gitma
Sangroniz 2
48150 Sondika
Vizcaya
Spain

www.gitma.es
Tel: +34 94-471-06-13
Fax: +34 94-453-61-21
E-mail: info@gitma.es

36, 114, 177, 178, 179, 198

Hess
Lantwattenstraße 22
78050 Villingen-Schwenningen
Germany

www.hess.eu
Tel: +49 7721-920-0
Fax: +49 7721-920-250
E-mail: hess@hess.eu

165

JCDecaux
17, rue Soyer
92 200 Neuilly-sur-Seine
France

www.jcdecaux.fr
Tel: +33 130-79-79-79
E-mail: info_ventes@jcdecaux.fr

38, 116, 163, 226

Kühn und Kirste

www.kuehn-und-kirste.de

68-69

Landscape Forms
431 Lawndale Ave.
Kalamazoo, MI 49048
USA

www.landscapeforms.com
Tel: +1 800-430-6209
Fax: +1 269-381-3455
E-mail: specify@landscapeforms.com

117, 216, 231

Larus
Vale da Mamõa,
3854-909 Albergaria-a-Velha
Portugal

www.larus.pt
Tel: +351 234-520-600
Fax: +351 234-520-609
E-mail: malmeida@larus.pt

40, 41, 42, 139, 218

Louis Poulsen
Kaistraße 20
40221 Düsseldorf
Germany

www.louispoulsen.com
Tel: +49 0211-73279-0
Fax: +49 0211-73279-100

39, 142, 143, 159

mago:urban

www.magogroup.com
Tel: +34 902 111 893
E-mail: info@magogroup.com

43, 44, 45, 46-47, 98-99, 112, 182, 183, 210, 211, 222

Metalarte
Tambor de Bruc 10
08970 San Joan Despí
Barcelona
Spain

www.metalarte.com
Tel: +34 93-477-00-69
Fax: +34 93-477-00-86
E-mail: metalarte@metalarte.com

144, 145, 169

Microarquitectura
Entença 218
08029 Barcelona
Spain

www.microarquitectura.com
Tel: +34 93-411-11-91
Fax: +34 93-491-20-97
E-mail: comercial@microarquitectura.com

232-233

miramondo
Paracelsusgasse 8/1/9
1030 Wien
Austria

www.miramondo.com
Tel: +43 1 96-90-404
Fax: +43 1 71-41-491
E-mail: office@miramondo.com

21, 48, 49

mmcité
Bílovice 519
687 12 Bílovice
Czech Republic

www.metalarte.com
Tel: +420 572 434 298
Fax: +420 572 434 283
E-mail: sales@mmcite.cz

50, 51, 52, 92-93, 118, 122, 123, 183, 220, 221, 228, 229, 230

Modo
Via Villafranca, 34
35010 - Campodoro - Padova
Italy

www.modoarredo.com
Tel: +39 049-906-5385
Fax: +39 049-906-5911

52, 53, 54, 88, 124, 125, 129, 184, 185, 196, 197, 226

Neri
S.S. Emilia 1622
47020 Longiano (FC)
Italy

www.neri.biz
Tel: + 39 0547-65-21-11
Fax: + 39 0547-54-074
E-mail: neri@neri.biz

55, 118, 188

Onn Outside
Ribera de erandio, 8
48950 Erandio Bizkaia
Spain

www.onnoutside.com
Tel: +34 94-417-10-30
Fax: +34 94-417-03-77
E-mail: info@onnoutside.com

20, 63, 106, 107, 199

Proiek Habita & Equipment
Pol. Bildosola Auzunea, J1.
48142 Artea
Bizkaia
Spain

www.proiek.com
Tel: +34 902-541-212
Fax: +34 902-331-902
E-mail: info@proiek.com

64-65

Runge
Postfach 36 46
49026 Osnabrück
Germany

www.runge-bank.de
Tel: +49 (0)541 50552-0
Fax: +49 (0)541 505522-2
E-mail: info@mail.runge.de

70, 215

Santa & Cole

www.santacole.com
Tel: +34 938-619-100
Fax: +34 938-711-767
E-mail: info@santacole.com

71, 72-73, 74, 119, 120, 130, 146, 147, 148, 149, 166, 167, 186, 189, 216, 217, 219

SIARQ
Montcada 31-33
08003 Barcelona
Spain

www.siarq.net
Tel: +34 93-55-33-913
Fax: +34 93-55-33-765
E-mail: infobcn@siarq.net

157

Siteco
Georg-Simon-Ohm-Straße 50
83301 Traunreut
Germany

www.siteco.com
Tel: +49 8669-33-0
Fax: +49 8669-33-397

150-151, 152-153

Street and Park Furniture
13/19 Heath Street, Lonsdale 5160
South Australia
Australia

www.streetandpark.com.au
Tel: +61 8 8329 6750
Fax: +61 8 8329 6799
E-mail: sales@streetandpark.com.au

75, 76, 97, 121, 158, 168, 181, 209

Streetlife
Oudesingel 144
NL-2312 RG
Leiden
The Netherlands

www.streetlife.nl
Tel: +31 (0)71-524-6846
Fax: +31 (0)71-524-6849
E-mail: streetlife@streetlife.nl

77, 78, 181, 194, 195, 207, 208, 222

T&D Cabanes
Parque Industrial Avanzado
Avenida de la Ciencia, 7
13005 Ciudad Real
Spain

www.tdcabanes.com
Tel: +34-926-25-13-54
Fax: +34 926-22-16-54
E-mail: info@tdcabanes.com

82, 83, 84, 119, 131, 154, 155, 156

Tecam BCN
Collita, parcela 11 isla 3
Pol. Ind Molí de la Bastida
08191 Rubí
Spain

www.tecambcn.com

90-91, 96, 187

VelopA
Beckerfelder Straße 96
47269 Duisburg
Germany

www.velopa.com
Tel: +49 (0)20-371-299-716
Fax: +49 (0)20-371-354-81
E-mail: info@velopa.com

80-81, 209

Westeifel Werke
Vulkanring 7
54568 Gerolstein
Germany

www.westeifel-werke.de
Tel: +49 0 6591-16-0
Fax: +49 0 6591-16-111

14-15, 86

Woodform
58 The Concourse, Henderson
PO Box 44-054, Point Chevalier
Auckland
New Zealand

www.woodbenders.co.nz
Tel: +64 9-835-4107
Fax: +64 9-835-4180
E-mail: dave@woodbenders.co.nz

Designer index

A
Aguado, Maria Luisa, 113
Albero, Roger, 46-47, 183
Albors, Eduardo, 154
Alday, Iñaki, 136
Ambasz, Emilio, 141
Área de diseño Gitma, 36, 114, 177, 178, 179, 198
Arola, Antoni, 144
Arribas, Alfredo, 148
Arriola, Andreu, 103
Artigues, Jaume, 204
Atelier Mendini, 16, 177

B
Barbato e Garzotto, 184
Basañez, Paúl, 64-65
Batlle, Enric, 71, 147, 166
BCQ Arquitectos, 87
Benedito, Ramón, 146
Bertólez, Guillermo, 120
Bolaños, Juan Carlos, 210
Brandt Dam, Erik, 115

C
Cabeza, Diana, 31, 32, 33, 85, 100, 101, 128, 227
Cabrera, Paula, 82, 156
Cabrera, Pere, 204
Canalda, Otto, 169
Cane, Brian, 216
Carandell, Joaquim, 137, 138, 175
Carvalho de Araújo, J.M., 40
Casamor, Carles, 113
Caviasca, Alessandro, 157
Chantilly design, 177, 178, 179, 198
Churtichaga, Jose María, 25
Cinca, Joan, 43
Cipollone, Eugenio, 22
Coleman-Davis Pagan Architects, 59
Cortella, Jean-Luc, 24
Cuenca Montilla, Juan, 119

D
Da Costa, Daciano, 40
de la Quadra-Salcedo, Cayetana, 25
de Marco, Antonio, 217
design-people, 39
díez+díez diseño, 102

E
Equipo de Durban Studio, 27, 28-29
Espinàs, Julià, 74

Estudio Hampton, 112

F
Farne, Alfredo, 118
Feduchi, Javier, 84
Fernández Castro, Roberto, 25
Fernández, Franc, 96
Fernández, Sergio, 90-91
Ferrándiz, Javier, 120
Ferraz, Marta, 82, 156
Figueras, Bet, 186
Fiol, Carme, 103
Foreign Office Architects, 98-99
Forgas, Joan, 12, 134
Fortunato, Diego, 109

G
Gabás, Maria, 113
Galán Lubascher, Lucas, 25
Galán Peña, José Luis, 25
Gaspar, Joan, 196
Ginés, Maurici, 136
Gómez-Pimienta, Bernardo, 13
González Cordón, Antonio, 149
Guimerà, Oriol, 167, 216

H
Häberli, Alfredo, 94-95
Healey, Greg, 121, 181, 209
Hegmon, Radek, 50, 92-93, 122, 123, 228, 229, 230
Helio, Piñón, 176
Herbut, Michelle, 97
Honkonen, Vesa, 160-161
Hruša & Pelcák Architects, 118

I
Irisarri - Piñeda, Jesús, 42

J
Jaques, Émanuelle, 211
Jarczak, Diego, 32, 128
Jover, Margarita, 136

K
Karásek, David, 50, 92-93, 122, 123, 228, 229, 230
Knudsen, Lauritz, 143

L
Lagranja, 219
Lauritzen, Vilhelm, 36
Licht Kunst Licht, 152-153

Lobo, Inês, 41
Lotersztain, Alexander, 60-61
Lovegrove, Ross, 200
Lozano, Alfredo, 84

M
Machimbarrena, Javier, 64-65
Mansilla+Tuñón, 18
Marforio, Enrico, 17, 114
Martínez Lapeña, J. A, 119
Matorell Pena, Bernard, 205
McCurry, Margaret, 117
Microarquitectura, 232-233
Milà, Gonzalo, 108
Milà, Miguel, 72-73, 108, 186
Morandi & Citterio, 185
Moreno, Pablo, 84
Muxart, Josep, 19, 89, 173, 192

N
Nahtrang Design, 56-57
Nouvel, Jean, 23

O
Odgård, Mads, 159
Outsign, 24, 111, 172

P
Pereira, Pedro, 218
Perera, Ernest, 211
Pericas, Enric, 89, 113
Periel, Montse, 189
Piaser, Alessandro, 88
PLH Design, 142
Providência, Francisco, 139, 218

Q
Quintana, Màrius, 232-233

R
Ramos / Bassols, 214
Ravaioli, Piero, 164
Regamey, Damien, 180
Rehwaldt Landschaftsarchitekten, 66-67, 68-69
Repossi, Tobia, 52, 124, 129, 184, 197
Riberti, Alessandro, 125
Riera Ubia, Ton, 44, 182
Rivoira y asociados, 112
Rodríquez, Francisco Javier, 62
Roig, Joan, 71, 147, 166

Rovero, Adrien, 223
Roviras, Pau, 130
Rubio, Germán, 26
Ruisanchez, Manuel, 58
Ruiz de Castañeda, Adolfo, 187
Ruiz, Mario, 145
Runge Designteam, 215

S
Sádaba, Juan, 20
Seste, 140
Siteco, 150-151
SLA, 156, 206
Smyth, Ted, 79
Soto, Vicente, 83
Starck, Philippe, 116, 163
Stella, Fausta, 54
Studio Italo Rota & Partners, 125
Studio MAO, 53
Studio Rota & Partners, 226
Suriñach, Josep, 35, 193
Szekely, Martin, 38

T
Tabuenca, Luis, 106
Tarrasó, Olga, 74
Thesevenhints, 21
Thygesen, Rud, 37
Torrente, Carlos, 130
Torres, Elías, 119

U
Úbeda, Ramón, 169
Urbanica, 111, 172

V
Valverde, Carles, 135
Valverde, Javier, 155
Vázquez Consuegra, Guillermo, 205
Venturotti, Alejandro, 32, 128
Viaplana, Albert, 176

W
Wehberg, Max, 14-15, 86
Winkels, Karsten, 165
Winkler, Bernhard, 30, 35, 110, 174, 198
Winkler, Thomas, 34
Wolfson, Martín, 100, 227